6G Wireless Communications and Mobile Networking

Edited by

Xianzhong Xie

School of Optoelectronic Engineering
Chongqing University of Posts and Telecommunications,
Chongqing, PR China

Bo Rong

Mikatel International Inc.,
Quebec, Canada

&

Michel Kadoch

École de Technologie Supérieure,
Université du Quebec,
Montreal, Quebec, Canada

6G Wireless Communications and Mobile Networking

Editors: Xianzhong Xie, Bo Rong, Michel Kadoch

ISBN (Online): 978-1-68108-796-2

ISBN (Print): 978-1-68108-797-9

ISBN (Paperback): 978-1-68108-798-6

need for a court order if at any point you breach any terms of this License Agreement. In no event will any delay or failure by Bentham Science Publishers in enforcing your compliance with this License Agreement constitute a waiver of any of its rights.

3. You acknowledge that you have read this License Agreement, and agree to be bound by its terms and conditions. To the extent that any other terms and conditions presented on any website of Bentham Science Publishers conflict with, or are inconsistent with, the terms and conditions set out in this License Agreement, you acknowledge that the terms and conditions set out in this License Agreement shall prevail.

Bentham Science Publishers Ltd.
Executive Suite Y - 2
PO Box 7917, Saif Zone
Sharjah, U.A.E.
Email: subscriptions@benthamscience.net

**BENTHAM
SCIENCE**

CONTENTS

PREFACE

Although 5G mobile networks have been standardized and deployed worldwide since 2020, the requirements of wireless communication services are not completely met, taking into account industrial and other challenging applications. Thus, the 6G wireless technologies kicked off the initial research and are expected to be applied around 2030. Different from 5G, the next generation networks highlight new features like high time and phase synchronization accuracy, near 100% geographical coverage, and high cost-efficiency. Compared with Gbps-level transmission data rate in 5G, a number of useful applications in 6G, such as high-quality 3D video, virtual reality (VR), and a mix of VR and augmented reality (AR), need Tbps-level transmission data rate that could be achieved with terahertz (THz) and optical technologies. Due to the big datasets generated by heterogeneous networks, the technology of artificial intelligence (AI) is regarded as a promising aid to wireless systems in a bid to improve the quality of service (QoS), quality of experience (QoE), security, fault management, and energy efficiency.

With the booming of higher frequency and more energy-saving equipment, THz and photonic communications become economically feasible. The advanced integrated circuit (IC) technology nowadays makes radio frequency (RF) devices and antennas more flexibly designed and highly integrated. Flexible RF components could work with artificial intelligence (AI) algorithms to make wireless networks more adaptive to user demand and RF environment. The 6G will undoubtedly expand the frequency from below 95 GHz to the high-frequency millimeter-wave and terahertz range. Those new frequency bands will focus on short-range communications by enabling the design of much tinier RF devices to support technologies like ultra-massive antenna arrays.

6G networks are envisioned to be full dimensional and would address every potential demand of services. A smart city is such a typical scenario where various Internet of things (IoT) applications proliferate to help citizen services. Other than conventional scenarios, smart city IoT services will rely on 6G networks for broad coverage, ultra-low latency, and reliable connection. As different IoT applications may have different service holders, it becomes necessary to employ network slicing (NS) to gain distinct virtual networks and differentiated quality of service guarantees. In the meantime, computing technologies such as cloud computing, fog computing, and edge computing are critical to network resilience, lower latency, and time synchronization. Cloud computing provides the ability to use flexible and telescopic services through various hosted services provided by the Internet. Edge computing, on the other hand, prefers the open platform using the network, computing, storage resources close to the site of the object in order to avoid the relatively long delay of accessing the cloud data center. Finally, big data technology can work with 6G to get hidden patterns, unknown relevance, potential trends, and other information.

The content of this book is summarized as follows.

1. In Chapter 1, we provide readers with a general vision of 6G, including the inevitability of 6G research, the international organizations for 6G standardization, and also the 6G research progress.

2. In Chapter 2, we introduce millimeter-wave technologies in 6G, including large-scale MIMO systems, precoding technology, and different kinds of beamforming structures. It also systematically summarizes the requirements on 6G millimeter-wave devices.

3. In Chapter 3, we focus on the development of the latest 6G antenna technology. In particular, it highlights the technical trends of a large-scale antenna from antenna design and synthesis to feed network and antenna selection.

4. In Chapter 4, we highlight the characteristics and application fields of the terahertz wave, especially the application in wireless communication. Two mainstream terahertz wireless communication systems are explained in detail under the context of 6G.

5. In Chapter 5, we propose a self-organizing network (SON) driven network slicing architecture, where software-defined networking (SDN) and network function virtualization (NFV) act as the key enablers. Some preliminary simulation results are given to validate the efficiency of the design.

6. In Chapter 6, we present an overview of the developing trend of IoT applications and discuss its relation to 6G. This chapter also sheds light on the challenges and solutions to future cellular massive IoT.

7. In Chapter 7, we give a systematic introduction to the cloud/edge computing and big data system in 6G. New applications, such as sensing, positioning, and slice-specific function, can significantly benefit from the new network computing architecture and AI-powered big data analysis.

Xianzhong Xie
School of Optoelectronic Engineering
Chongqing University of Posts and Telecommunications
Chongqing, PR China

Bo Rong
Mikatel International Inc.
Quebec, Canada

&

Michel Kadoch
École de Technologie Supérieure,
Université du Quebec,
Montreal, Quebec, Canada

List of Contributors

Aart W. Kleyn   Center of Interface Dynamics for Sustainability, Institute of Materials, China Academy of Engineering Physics, Chengdu 610200, Sichuan, China

Anthony J. Vickers   Department of Electronic Systems Engineering, University of Essex, Wivenhoe Park, Colchester, Essex, UK

Brian D. Gerardot   Institute of Photonics and Quantum Sciences, SUPA, Heriot-Watt University, Edinburgh, EH14 4AS, UK

Bo Rong   Mikatel International inc., Quebec, Canada

Bo Yin   College of Electronic Engineering, Chongqing University of Posts and Telecommunications, Chongqing, China

Fanqin Zhou   State Key Laboratory of Networking and Switching Technology, Beijing University of Posts and Telecommunications, Beijing, P. R. China

Fei Qi   China Telecom Beijing Research Institute, Beijing, China

Jia Ran   College of Electronic Engineering, Chongqing University of Posts and Telecommunications, Chongqing, China

Lei Shi   Carlow Institute of Technology, Carlow, Ireland

Michel Kadoch   ÉÉcole de Technologie Supérieure, Université du Quebec, Montreal, Quebec, Canada

Min Wang   College of Electronic Engineering, Chongqing University of Posts and Telecommunications, Chongqing, China

Mohamed Cheriet   École de Technologie Supérieure, Université du Quebec, Montreal, Canada

Peng Yu   State Key Laboratory of Networking and Switching Technology, Beijing University of Posts and Telecommunications, Beijing, P. R. China

Tao Hong   School of Electronic and Information Engineering, Beihang University, Beijing, China

Wei Luo   College of Electronic Engineering, Chongqing University of Posts and Telecommunications, Chongqing, China

Xianzhong Xie   School of Optoelectronic Engineering, Chongqing University of Posts and Telecommunications, Chongqing, P. R. China

CHAPTER 1

Explaining 6G Spectrum THz, mmWave, Sub 6, and Low-Band

Xianzhong Xie[1,*], **Bo Rong**[2] and **Michel Kadoch**[3]

[1] *School of Optoelectronic Engineering Chongqing University of Posts and Telecommunications, Chongqing, P.R. China*

[2] *Mikatel International Inc., Quebec, Canada*

[3] *École de Technologie Supérieure, Université du Quebec, Montreal, Quebec, Canada*

Abstract: This chapter aims to provide readers with a general vision of 6G. Firstly, we give a simple overview of various aspects related to 6G, including inevitability of 6G research, international organizations for standardization, and also 6G research progress of some countries/regions. Then, 6G spectrum compositions are discussed in detail with emphasis on SUB-6, mmWAVE, and Terahertz (THz).

Keywords: 6G, Frequency spectrum.

THE SIXTH GENERATION MOBILE COMMUNICATION (6G)

The development of mobile/wireless communication has gone through the process of 1G/2G/3G/4G, and it has entered a critical stage of 5G commercial development. From the historical perspective of industrial development, the mobile communication system has been updated every ten years. The increasing demand for user communication and the innovation of communication technology is the driving force for the development of mobile communication [1]. However, 5G will not meet all requirements of the future of 2030 and beyond [2]. Researchers now start to focus on the sixth-generation mobile communication (6G) networks. Some countries and organizations have already initiated the exploration of 6G technology with the launch of 5G commercial deployment in major countries around the world.

* **Corresponding author Xianzhong Xie:** School of Optoelectronic EngineeringChongqing University of Posts and Telecommunications,Chongqing, P.R. China; Tel: 0086 23 62460522; Fax: 0086 23 62471719; E-mail: xiexzh@cqupt.edu.cn

THE INEVITABILITY OF 6G RESEARCH

The 10-year Cycled Rule

Since the introduction of the first generation (1G) mobile communication system in 1982, a new generation of wireless mobile communication systems has been updated approximately every 10 years, as shown in Fig. (**1**). It will take about 10 years from conceptual research to commercial applications [3]. In other words, when the previous generation enters the commercial period, the next generation begins conceptual and technical research. 5G research started 10 years ago, and now 6G research is in line with the development law of mobile communication systems. It may take about ten years for 6G to arrive, but research on 6G cannot be delayed. Mobile communications will stride towards the 6G era.

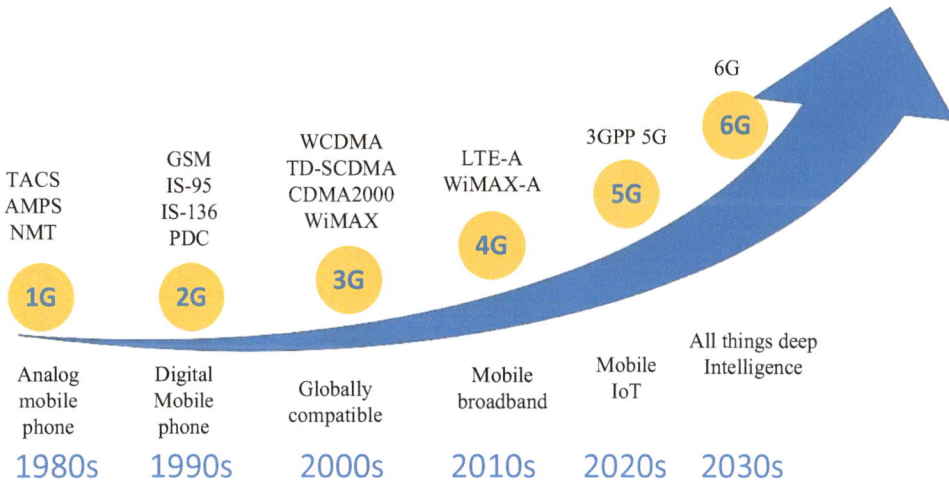

Fig. (1). The evolution of mobile communication systems.

"Catfish Effect"

The "catfish effect" means that it also activates the survival ability of the small fish when the catfish disturbs the living environment of the small fish. It is to adopt a means or measures to stimulate some enterprises to become active and invest in the market to actively participate in the competition, which will activate enterprises in the same industry in the market. 5G is different from previous generations of mobile communication systems mainly aimed at IoT/vertical industry application scenarios. Many vertical industry members will definitely participate in the 5G ecosystem with the large-scale deployment of 5G networks. The in-depth participation of emerging companies (especially internet companies born with innovative thinking) in the future will have a huge impact on the traditional communications industry and even a revolutionary impact compared with the status quo dominated by traditional operators, which is called "catfish effect".

The Explosive Potential of IoT Business Models

IoT is the inevitability of the internet from top to bottom in the industry. It is an extension from the inside out, with the cloud platform as the center. Just as the emergence of smartphones stimulated 3G applications and triggered the demand for large-scale deployment of 4G, it is believed that certain IoT business models will also stimulate the 5G industry to burst at a certain point in the 5G era, which will stimulate the future needs of 6G networks. To accommodate the stringent requirements of their prospective applications, we need to have enough imagination. We must prepare in advance for the possible future network and lay a good technical foundation [4]. Based on the above analysis, we can draw the conclusion that now is the right time to start the research on the next generation wireless mobile communication system.

The 5G Performance Would Limit New IoT Applications

Despite the strong belief that 5G will support the basic MTC and URLLC related IoT applications, it is arguable whether the capabilities of 5G systems will succeed in keeping pace with the rapid proliferation of ultimately new IoT applications [5]. Meanwhile, following the revolutionary changes in the individual and societal trends, in addition to the noticeable advancement in human-machine interaction technologies, the market demands by 2030 are envisaged to witness the penetration of a new spectrum of IoT services. These services deliver ultrahigh reliability, extremely high data rates, and ultralow latency simultaneously over uplink and downlink [6]. The unprecedented requirements imposed by these services will push the performance of 5G systems to its limits within 10 years of its launch. Moreover, these services have urged that 6G should be capable of unleashing the full potentials of abundant autonomous services comprising past as well as emerging trends.

INTERNATIONAL ORGANIZATION FOR STANDARDIZATION

International Telecommunication Union (ITU)

According to the ITU work plan, the RA-19 meeting in 2019 will not establish a new IMT technical research resolution. It indicates that the research cycle from 2019 to 2023 is still mainly for 5G and B5G technology research, but the 6G vision and technology trend research will be carried out from 2020 to 2023. The mainstream companies in the industry generally believe that it is more appropriate to establish the next generation IMT technology research and naming resolution at the RA-23 meeting in 2023. ITU-T SG13 (International Telecommunication Union Telecommunication Study Group 13) established the ITU-T Focus Group Technologies for Network 2030 (FG NET-2030) at its meeting in July 2018. The FG NET-2030 intends to define the requirements of networks of the year 2030 and beyond [7]. 6G research work was carried out at ITU-R WP5D (the 34th International Telecommunication Union Radio Communication Sector 5D Working Group meeting) held in February 2020, which includes the formulation of a 6G research timetable, future technology trend research reports, and the writing of future technology vision proposals.

The Third Generation Partnership Project (3GPP)

3GPP is the main promoter and integrator of communication system technical specifications managing the standardization work such as the introduction of communication system requirements, system architecture design, security, and network management. 3GPP has completed the development of the first version of the 5G international standard Release 15 (R15) focusing on supporting enhanced mobile broadband scenarios and ultra-high reliability and low-latency scenarios in June 2018. The development of a complete 5G international standard Release 16 (R16) will be completed in Autumn 2020, which will fully support the three application scenarios determined by the ITU. Also, 3GPP is promoting 6G research and standardization activities. 3GPP Release 17 (R17) has started to investigate advanced features that would shape the evolution toward 6G [8], and substantive 6G international standardization is expected to start in 2025.

Institute of Electrical and Electronics Engineers (IEEE)

To better summarize and sort out the related technologies of next-generation networks, IEEE launched the IEEE 5G Initiative in December 2016 and renamed it IEEE Future Networks in August 2018 to enable 5G and the future network. IEEE is also developing corresponding 5G standards, which are expected to be submitted to ITU for approval in 2020. At present, IEEE has carried out some 6Gtechnical seminars. The 6G wireless summit will be held by IEEE in March every year, and the first 6G wireless summit was initiated by IEEE in the Netherlands on March 25, 2019. The industry and academia were invited to publish the latest insights on 6G. The theoretical and practical challenges that need to be addressed to realize the 6G vision are discussed. The global 6G research vision, requirements and potential approaches were published in 6G White Papers at the end of June 2020.

6G RESEARCH PROGRESS IN SOME COUNTRIES/REGIONS

6G communication is still in its infancy. The 6Gresearch race from academia can be said to have started in March 2019, when the first 6G Wireless Summit was held in Levi, Finland. Some researchers also defined 6G as B5G or 5G+. Preliminary research activities have already started in some countries/regions. The US president has requested the deployment of 6G in the country. China has already started the concept study for the development and standardization of 6G communications in 2019. Most European countries, Japan, and Korea are planning several 6G projects.

European Union

The European Union initiated consultation for 6G technology research and development projects aiming to study key technologies for next generation mobile communications in 2017. Finnish 6G research activity is coordinated by the University of Oulu in 2018, where a 6G initiative was launched. The EU's preliminary assumptions for 6G are that the peak rate should be greater than 100 Gbit/s, the single channel bandwidth can reach 1 GHz and the terahertz frequency band higher than 275 GHz should be used. The European Union launched a three-year 6G basic technology research project in 2019. The main task is to study the next generation error correction coding, advanced channel coding and modulation technologies used in 6G networks. Besides, the EU has also initiated a number of terahertz research and development projects. The EU has listed the development of terahertz communications as a 6G research program. A research group based on the EU's Terranova project is now working toward the reliable 6G connection with 400 Gbit per second transmission capability in the terahertz spectrum.

United States

The FCC (The United States Federal Communications Commission) launched CBRSD (Public Wireless Broadband Service) in the 3.5GHz frequency band in 2015, which dynamically manages different types of wireless traffic through a centralized spectrum access database system to improve spectrum utilization efficiency. The experts from FCC proposed three key 6G technologies at the "Mobile World Congress 2018-North America" summit in September 2018, including new spectrum (terahertz frequency band), large-scale spatial multiplexing technology (supporting data hundreds of ultra-narrow beams) and blockchain based dynamic spectrum sharing technology. In addition, FCC announced that it would open the terahertz frequency band (95 GHz-3 THz) for using in 6G technology trials in March 2019, thereby setting the US as the pacesetter in the 6G research race.

Japan

Nihon Keizai Shimbun reported that Japan's NTT group has successfully developed new technologies for B5G and 6G in July 2018. One is Orbital Angular Momentum (OAM) technology. It has realized the superimposed transmission of 11 radio waves that are several times of 5G. OAM technology uses a circular antenna to rotate radio waves into a spiral for transmission. Due to the physical characteristics, the high number of revolutions will make the transmission more difficult. NTT plans to realize the superposition of 40 radio waves in the future. And the other is terahertz communication technology. The development of terahertz technology is listed as the top of the "ten key strategic goals of national pillar technologies". And a budget of more than 1 billion yen is proposed in the fiscal year 2019 to start research on 6G technology. The peak transmission rate reaches up to 100 Gbit/s. Japan still faces the problem of extremely short transmission distances, but the transmission speed can reach 5 times that of 5G. Japan readies US$2 billion to support industry research on6G technology. NTT and Intel have decided to form a partnership to work on 6Gmobile network technology. In addition, an EU–Japan project under Horizon2020 ICT-09-2017 funding called "Networking Research beyond 5G" also investigated the possibility of using the THz spectrum from 100 to 450 GHz.

South Korea

Experts from SK Telecom's ICT R&D Center presented 3 technologies for future 6G networks at a cutting-edge technology seminar held at New York University in October 2018, including terahertz communications and de-cellular architecture (fully virtualized RAN+ large-scale antennas) and non-terrestrial wireless networks. Samsung Electronics and SK Telecom work together to develop technologies and business models related to 6G. LG Electronics established a 6G research center in collaboration with the Korea Advanced Institute of Science and Technology. In addition, SK telecom has reached an agreement with two equipment manufacturers, Ericsson and Nokia, to jointly develop 6G technologies. Korean operators achieved download speeds of 193~430 Mbps in the 3.5 GHz (sub-6GHz) frequency band in April 2019.

China

China began to study the 6G mobile communication system to meet the inconstant and rich demands of the IoT in the future at the end of 2017, such as medical imaging, augmented reality and sensing. In addition to solving the problems of wireless communication between people and wireless internet access, it is also necessary to solve the communication between things and things and between people and things with the expansion of the use of mobile communication in the future. 6G communication technology mainly promotes the development of the IoT. The Ministry of Science and Technology of China (MSTC) declared its goal of leading the wireless communication market in the 2030s by expanding research investment in 6G, and issued a notice on the annual project application guidelines for key special projects in 2018 such as "Broadband Communications and New Networks" for the national key research and development plan, 5 of which involve B5G/6G. In 2019 the MSTC also planned to set up two working groups to carry out the 6G research activities: the first is from government departments to promote 6G research and development, the second is made up of 37 universities, research institutes and companies, focusing on the technical side of 6G.

6G SPECTRUM COMPOSITION

Spectrum Requirements for 6G

Our society will become data-led due to almost no time-delay wireless connections by around 2030. Therefore, 6G will be expected to be used to promote the development of the wireless technology we are familiar with today. And it will be expected to achieve a quite good system performance. Fig. (**2**) presents a synopsis of the evolving wireless cellular communication generation. Specifically, frequency spectrum used by various generations of mobile communication systems is shown. In order to increase the data rate 100 to 1000 times faster than 5G in terms of frequency spectrum [9], 6G may use a higher frequency spectrum than previous generations as a vision for the future.

4G and the previous mobile communication systems all use the Sub-6 GHz frequency band, while the 5G mobile communication system uses both the Sub-6 GHz frequency band (FR1 band, 450 MHz-6 GHz) and the 24-100 GHz frequency band (FR2 band, 24.25 GHz-52.6 GHz) [10]. Researchers realize that although 5G expands the spectrum bandwidth, the current frequency band is still not enough to meet the increasing demand for communication services in the rapidly developing human society. Therefore, in the study of 6G networks, we will consider spectrum resources above 100 GHz such as millimeter wave (mmWave) and terahertz (THz) to increase the transmission bandwidth.

The 6G wireless communication system will use multi-band and high-spread spectrum to increase the transmission rate. The ultimate vision is to make the end-to-end transmission rate reach hundreds of gigabits. It is expected that in the future 6G, the ground mobile communication network, satellite system, and Internet will merge into a large space-air-ground-sea network. Thus, 6G spectrum needs to support space-air-ground-sea integrated communications. Most operating frequencies for space-air-ground-sea communications are assigned by the

International Telecommunication Union (ITU) [11]. The mmWave bands can be used in both space-ground and air-ground channels as well as space-air transmission.

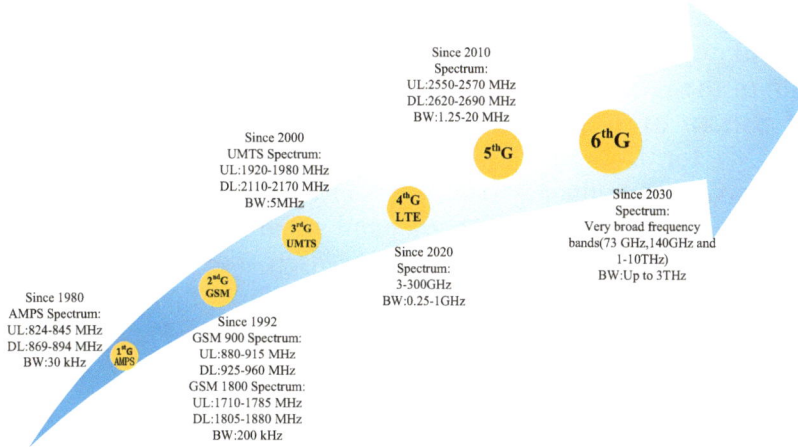

Fig. (2).   Spectrum compositions of mobile communication systems.

SUB-6

Sub-6 GHz bands have been the primary working frequency in the third generation, 4G, and 5G due to wide coverage capabilities and low cost, which are also indispensable in 6G [12].

The Low Frequency Spectrum

Added Spectrum of 6 GHz

As the saying goes, "A single wire cannot form a thread, and a single tree cannot form a forest." The development of 5G not only requires a large amount of spectrum resources, but also requires high, medium and low frequency collaborative work. The World Radio Communication Conference in 2019 (WRC-19) reached a global consensus on the 5G millimeter wave frequency band to meet the business needs of 5G systems for ultra-large capacity and high-speed transmission. At the same time, in order to solve the problem of large-scale and deep coverage for 5G systems and achieve a better balance between network capacity and coverage, many countries are focusing on continuous 5G spectrum in the middle and low frequency bands. The new IMT (5G or 6G) usage rules for the 6GHz (5925 MHz-7125MHz) frequency band was included in the agenda of WRC-23 with the vigorous promotion of the Chinese delegation at the WCR-19.

6425 MHz-7025 MHz becoming a new regional (Arab countries, Africa, Europe, CIS countries) IMT frequency band and 7025 MHz-7125 MHz becoming a new global IMT frequency band are being investigated.

Spectrum resources are precious and scarce as the core resource for the development of mobile communication technology. Spectrum planning is the starting point of the industry and will also determine the development direction, rhythm and pattern of the industry. The successful establishment of the new allocation of IMT for the 6 GHz spectrum means that the 6 GHz frequency band will become a potential frequency band for IMT (5G or 6G).Countries around the world will give priority to this frequency band when they build 5G systems and future 6G systems, which promotes the research and development of IMT technology for 6 GHz band and the internationalization of the industrial chain and accelerates the process of 5G global commercial and 6G research and development.

Capacity and Coverage

5G new business applications have driven rapidly increasing in mobile data usage with the further acceleration of 5G commercial use. And enhanced mobile broadband services, fixed wireless broadband services and industrial applications such as smart cities and industrial manufacturing have accelerated the surge in mobile data usage. According to an industry analysis report, data usage per user per month in some leading markets will reach 150 GB in 2025. This requires a large amount of radio spectrum to support undoubtedly. The low and medium frequency bands can provide continuous bandwidth on the order of 100 MHz and good network coverage compared with the millimeter wave frequency band. Furthermore, the performance requirements of network capacity and coverage can be taken into consideration, and network construction costs can be greatly reduced. In addition, the propagation characterization and channel models in Sub-6 GHz bands have been extensively investigated in 5G. Therefore, they are important parts of the 5G or 6G spectrum. As one of the pioneers in the development and deployment of 5G systems, China's radio management department has been committed to seeking more IMT frequency resources for 5G or future 6G technologies to support its future technologies and applications from the perspective of efficient frequency use and long-term planning.

According to the characteristics of each frequency band, the Sub-6 GHz spectrum will take into account the requirements of coverage and capacity, which is an ideal compromise between peak rate and coverage capability. The frequency spectrum above 6GHz can provide ultra-large bandwidth, larger capacity and higher speed, but the continuous coverage capability is insufficiently shown in the Fig. (**3**).

5G wireless infrastructures for Sub-6 GHz will be widely deployed using a beamforming solution, which can greatly expand network coverage and penetration capabilities within buildings. Low frequency bands (such as 600 MHz, 700 MHz, 800 MHz, 900 MHz, 1.5 GHz, 2.1 GHz, 2.3 GHz and 2.6 GHz) have the characteristics of wide coverage and low cost, which are used the large-scale IoT in the future, the industrial automation and the key task IoT. Wireless throughput and capacity will show explosive growth with the mobile networks continuing to accelerate. The above 24GHz mmWave beamforming is also considered as a promising technology to provide ultra-high capacity in 6G coverage.

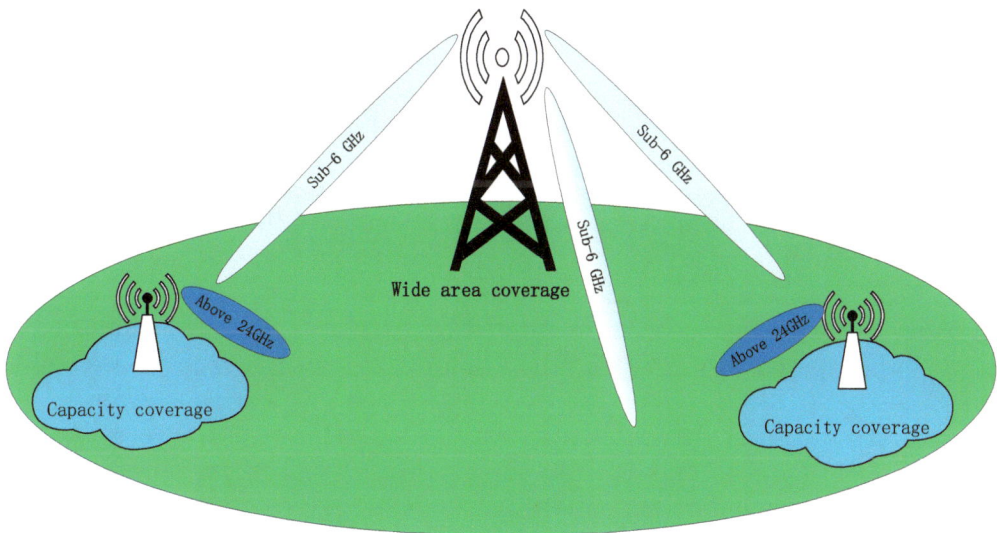

Fig. (3). Coverage of frequency bands.

The Spectrum Allocation of Sub-6

Spectrum for 5G NR

The spectrum resource from 1.7 GHz to 4.7 GHz is the Sub-6 GHz spectrum allocated to 5G NR by 3GPP, which is the FR1 band, and the maximum continuously allocated spectrum is 100 MHz [13]. The following is an introduction to several major FR1 bands. n77 (3300 MHz-4200 MHz) and n78 (3300 MHz-3800 MHz) are currently the most unified frequency bands for 5G

NR in the world. n79 (4400 MHz-5000 MHz) is also used in 5G NR mainly promoted by China, Russia and Japan. n28 (700 MHz) is also highly valued because of its good coverage. This frequency band has been identified as a pioneer candidate frequency band for global mobile communications at WRC-15. If this frequency band cannot be fully utilized, it would be a pity. At present, the US operator T-Mobile has announced the use of n71 (600 MHz) to build 5G. The use of the 3.4 GHz-3.8 GHz band areas is the most complicated [14]. For example, in the United States, the 3.5 GHz frequency band is used for Citizen Broadcast Radio Service (CBRS), while 150 MHz of the frequency spectrum is used for radar communications. In addition, the frequency band can also be used for other commercial services using dynamic access [15]. This dynamic access method saves users from expensive spectrum licenses, only needs to pay the corresponding communication fees to the service provider [16].

Spectrum Selection of Systems

License-free frequency band technology simplifies the process and restrictions for users to connect to the network. Nevertheless, due to the existence of these unlicensed spectrum users in the network, the reliability and security of the wireless network will be disturbed.

In the Sub-6 GHz spectrum area, the most noteworthy unlicensed spectrums are the 2.4 GHz and 5 GHz bands, which are currently the two frequency bands with the highest utilization rate. Among them, bandwidth resources in the 2.4 GHz band are scarce and have been preempted by existing services, making it very crowded. Compared with 2.4 GHz, the available frequency of the 5 GHz band is wider. But with the development of wireless communication networks, it is expected that the 5 GHz band will also be taken up in the next few years. And the current 5 GHz band is divided into several parts by different wireless access methods, which cannot be used uniformly.

Because the spectrum resources used by wireless communication networks and radar systems overlap, the Dynamic Frequency Selection (DFS) mechanism is required by the FCC and ETSI to be used in the Sub-6 GHz band to avoid interference to the radar system. And most of the access methods for the available spectrum in the low-frequency band are the Listen Before Talk (LBT). LBT will cause a large amount of delay idle periods in the use of spectrum, resulting in low spectrum utilization, which is the major resistance to the low-latency vision of 5G networks.

Recently, FCC has promoted additional spectrum for unlicensed usage in the 5.925 GHz - 7.125 GHz range. This is commonly referred to the 6 GHz band, and its regulations are currently been defined. The current trend of extending the availability of unlicensed access, even in the below 10 GHz spectrum region, copes with the necessity of dealing with the spectrum crunch due to the exponential increase of wireless applications [17]. Unlicensed access also eliminated the obsolete licensing paradigm that is known to lead to inefficient spectrum utilization.

mmWAVE

It is known that the target data rate of Sub-6 GHz 5G mobile communications is Gbps level, and the target data rate of 5G mmWave is about 10 Gbps. There are two key ways to increase the wireless transmission data rate: one is by improving the spectral efficiency and the other is by using large frequency bandwidth or spectrum resource [18].

6G mmWave Communication

Advantages of mmWave

In the development of mobile communication systems in the past, low-frequency microwave communications represented by 2.4 GHz and 5 GHz bands have been fully studied. However, as a result of its limited physical frequency width and being occupied by more and more network traffic, low-frequency microwave communication has become very crowded and cannot support the capacity requirements of a new generation of mobile communication systems [19].

Therefore, millimeter wave (mmWave) communication will become one of the main supporting technologies of a new generation of wireless communication systems by virtue of its large bandwidth, low latency, high rate, and friendly positioning function. The mmWave band is usually defined as electromagnetic waves in the 30GHz-300GHz frequency domain. It is located in the overlapped wavelength range of microwave and far-infrared waves, so it has the characteristics of both spectrums. Since its wavelength is 1ms-10ms, it is called the millimeter wave.

Since the development of the Internet, the democratization of information has improved education, e-commerce has promoted economic growth, and business innovation has been accelerated by supporting broader cooperation. And now we are entering an era of the Internet of Everything (IoE) integrating billions or even trillions of connections. Everything will gain contextual awareness, enhanced processing capabilities, and better sensing capabilities. Therefore, The IoE will become an application scenario that must be considered in the new generation of mobile communication systems. The large bandwidth and high speed that mmWave can bring make it possible for businesses such as high-definition video, virtual reality, augmented reality, dense urban information services, factory automation control, and telemedicine. It can be said that the development and utilization of mmWave provides a broad space for 5G applications on the basis of Sub-6 services.

But there is a huge disadvantage that mmWave has serious attenuation during propagation compared with other frequency bands [20]. And because of its large available bandwidth, the original resolution of mmWave communication systems is smaller than that of traditional low-band communication systems [21, 22]. Therefore, mmWave has not appeared in the field of mobile communication for a long time. Qualcomm prescribes the right medicine. Firstly, it uses multi-beam technology to address mobility challenges and improve coverage, robustness and non-line-of-sight operation. Secondly, it uses path diversity to deal with blocking problems, and uses terminal antenna diversity to improve reliability. Finally, several indoor and outdoor OTA tests were carried out to verify mobility. Therefore, the application of mmWave in mobile phones is realized.

Unlicensed mmWave Bands

In the 60GHz mmWave frequency band, about 7 GHz bandwidth is available for unlicensed access, which is much larger than the unlicensed spectrum of sun-6GHz. In such a wide available spectrum, to achieve the peak transmission rate of 100 Gbps required by the 5G network, at least a spectrum utilization rate of 14 bit/Hz per unit time is required. However, the existing data modulation/demodulation technology and related original devices cannot meet such strict spectrum utilization and symbol fidelity. Consequently, one of the current research difficulties in the field of wireless communication systems is how to achieve a transmission rate of 100 Gbps in a high-frequency band above 100 GHz [23]. For example, the FCC decided to open up four mmWave frequency bands above 95 GHz in the United States in 2019. As shown in Table **1**, a total of approximately 21 GHz bandwidth is provided for license-free access.

Spectrum Options for 6G

The 26 GHz-100 MHz frequency band is a big opportunity for mobile networks. However, it is the previously unpopular feature of mmWave that makes it very suitable for 5G. It relies on the use of micro-infrastructure, such as small units distributed in dense urban locations. 5G needs to use a wider spectrum than previous generations of mobile systems. According to this trend, it is expected that a higher frequency spectrum will be used in 6G wireless communication networks. Therefore, we initially set a 6G system to be deployed in the subterahertz band of 114 GHz-300 GHz as shown in Fig. (**4**). According to its ultra-high frequency physical characteristics, the sub-terahertz spectrum can complete ultra-narrow beam transmission, thereby improving spectrum efficiency and performing precise positioning [24]. This allows the sub-terahertz band to perform well in scenarios such as future backhaul networks and real-time short-distance communications.

Table 1. Unlicensed mmWave bands.

Frequency Band (GHz)	Contiguous Bandwidth (GHz)
116-123	7
174.8-182	7.2
185-190	5
244-246	2
Total	21.2

On the other hand, It is predicted that the number of mobile devices worldwide will be more than 125 billion in the 2030s, including the mobile phone, tablet, wearable devices, integrated headsets, implantable sensors and other machine-type users. The Internet of Thing (IoT) system, connecting millions of people and billions of devices, will be one of the essential application scenarios for mmWave B5G and 6G.

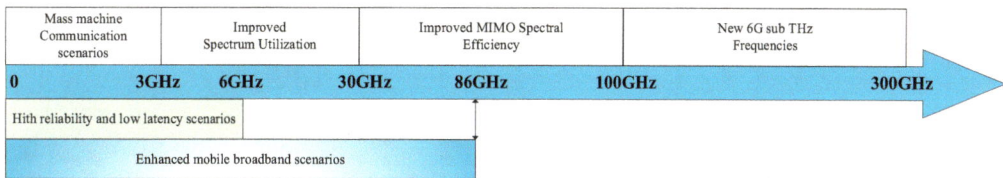

Fig. (4). Spectrum options for 6G.

Terahertz (THz)

As the propagation medium of wireless mobile communication systems, spectrum resources will become increasingly scarce as the types and demands of communication services continue to develop. At the beginning of 5G research and development, there have been suggestions to seek scalable spectrum resources in the direction of terahertz and visible light.

Terahertz is a new radiation source with many unique advantages. It has a wide range of applications in many fields, such as semiconductor materials, tomography technology, label-free genetic examination, broadband communications, and microwave orientation. The study of radiation sources in this frequency band will not only promote the major development of theoretical research but also pose major challenges to solid-state electronics and circuit technology. The development and utilization of terahertz spectrum in fields such as communications have been highly valued by Europe, the United States, Japan and other countries and regions. And they have also received strong support from the ITU [25].

6G Terahertz Communication

Terahertz Spectrum

With the development of science and technology, the communication volume and connection volume of wireless services are increasing explosively. Especially since the 5G era, the rapid development of wireless communication technologies in space and sky fields has made 2-dimensional wireless channel performance analysis no longer applicable. In the 6G network, we need to perform three-dimensional modeling of the wireless channel and calculate the relevant performance parameters in the unit cubic space, which requires a wider RF spectrum bandwidth. The frequency spectrum is a physical quantity existing in nature, which cannot be increased or decreased, so it is extremely precious. Terahertz waves refer to electromagnetic waves with frequencies in the range of 0.1THz-10 THz, between millimeter waves and infrared light as shown in Fig. (**5**). It is the transition zone from macroscopic classical theory to microscopic quantum theory, as well as the transition zone from electronics to photonics, called the terahertz gap of the electromagnetic spectrum.

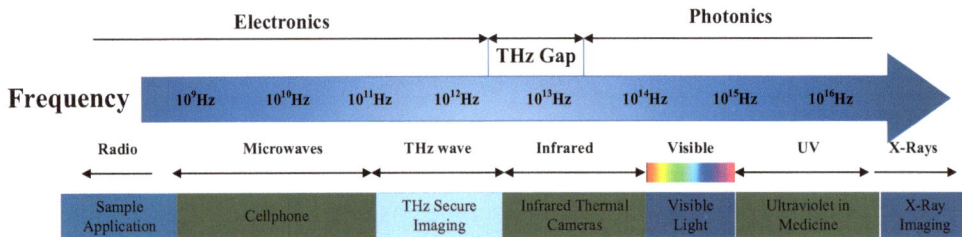

Fig. (5). Spectrum of mmWave and terahertz.

The commercial application of terahertz technology did not seem clear enough a few years ago. However, the terahertz measuring instruments showed great market potential in 2018. Civil safety applications, non-destructive testing and industrial quality control can all benefit from the application of the terahertz systems. Traditionally, the 26.5–300 GHz frequency band is defined as the mmWave band and 300–10000 GHz as the THz band. In recent years, however, the definition of 100–10000 GHz (or 0.1–10 THz) as the THz band has been commonly accepted.

With the critical growth of wireless communication, THz frequency bands have been envisioned as a promising candidate to supply sufficient spectrum resources for near future 6G. Specifically, the terahertz band can be up to four orders of magnitude higher than the existing microwave communication, which means that it can carry a larger amount of information. Also, terahertz integrates the advantages of microwave communication and optical communication, with many characteristics such as high transmission rate, large capacity, strong directionality, high security, and good penetration [26]. Enabled by these obvious advantages, extensive researches of THz communication have been developed worldwide.

Advantages of Terahertz

Terahertz communication has many advantages over microwave and wireless optical communication with its unique characteristics. With the development of terahertz technology, its unique advantages, and huge application prospects in many important fields of communication radar, deep-space communication, high-speed short-range wireless communication, and security wireless communication are gradually revealed [27, 28].

1. Terahertz wave is more suitable for high-speed short-range wireless communication because it is easily affected by moisture in the air when it propagates, causing serious signal attenuation.
2. Terahertz waves have narrow beams, good directivity and strong anti-interference capabilities. Therefore, secure communication within 2-5 km can be realized using terahertz waves.
3. The high frequency and wide bandwidth physical characteristics of terahertz can meet the ever-increasing demand for wireless communication capacity. The terahertz spectrum has a usable spectrum bandwidth of tens of GHz between 108-1013 GHz, which can provide a rate exceeding Tbit per second.
4. In terms of space communications, there are relatively transparent atmospheric windows near the 350, 450, 620, 735 and 870 μm wavelengths of terahertz waves shown in Fig. (6), which can achieve lossless transmission. Very small power can complete long-distance communication. Moreover, the receiving

terminal is easy to align because the terahertz wave beam is wider compared with wireless optical communication. The quantum noise is lower and the antenna can be miniaturized and planar. Therefore, terahertz waves can be widely used in space communications and are particularly suitable for wide-band communications such as satellite communications and satellite-to-ground communications.

5. The terahertz band is also suitable for Massive MIMO [16] with more antennas (the same size or even smaller antenna volume compared to millimeter waves) because of a short wavelength. Preliminary research shows that the beamforming and spatial multiplexing gain provided by Massive MIMO can well overcome the rain fading and atmospheric fading of terahertz propagation, which can meet the coverage requirements of dense urban areas such as the cell with a radius of 200 m.

6. Due to the low photon energy of terahertz waves, compared with wireless optical communication, terahertz waves as information carriers have the advantage of high energy efficiency.

7. Terahertz has a very short wavelength, has a high time-domain spectrum signal-to-noise ratio, can penetrate walls with a small attenuation, and is an ideal technology for imaging through walls in complex environments.

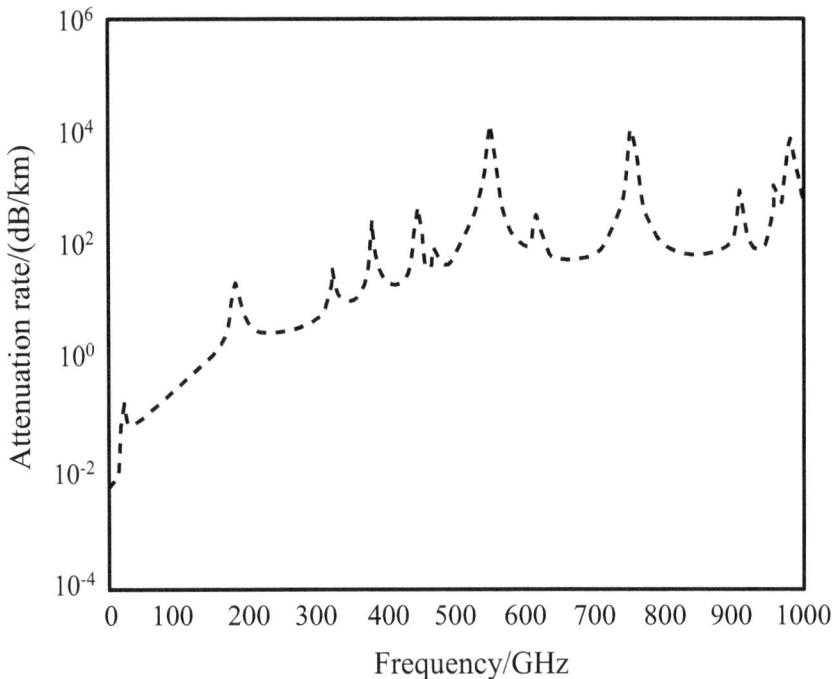

Fig. (6). The attenuation rate of terahertz.

Challenges in the Terahertz Bands

The terahertz frequency band has irreplaceable benefits for mobile communication. However, it also faces many challenges [29, 30].

1. Communication distance: Due to the short wavelength and high frequency of terahertz, it has a larger free space fading than low-frequency bands. In addition, the existing communication mechanisms may not be suitable for the highly directional propagation characteristics of terahertz.
2. Large-scale fading: Shadows have a non-negligible effect on the transmission of terahertz radio signals with ultra-high frequencies. For example, after a terahertz signal passes through the human body, its signal is attenuated by at least 20 dB, and after passing through a brick wall, its signal attenuation can even be as high as 80 dB. However, because humidity/rainfall fading is mainly for radio signals with frequencies sub-100 GHz, the terahertz band above 100 GHz has less impact.
3. Fast channel switching and disconnection: The channel coherence time has a linear relationship with the carrier frequency for a given moving speed. Therefore, because the coherence time is very small, terahertz has a greater Doppler shift, and higher shadow fading leads to significant path loss, which ultimately makes terahertz communication have the problems of frequent frequency band switching and severe fluctuate. Also, terahertz technology is mainly used for covering high-directional communication in a small area. This signifies that path fading, serving beams, and cell association relationships will change rapidly. To solve the problem of connection interruption caused by frequent frequency switching, we need to design a new rapid adaptive communication mechanism from a system perspective.
4. Energy consumption: A major difficulty in using ultra-large-scale antennas is the energy consumption of broadband terahertz system analog/digital conversion. Power consumption generally has a linear relationship with the sampling rate and an exponential relationship with the number of samples per unit. The wide bandwidth of the terahertz frequency band and the large number of antennas require high-resolution quantification. Therefore, it will be a huge challenge to implement low-power and low-cost devices.

Related Technology for Terahertz Communication

The following are several key research contents involved in the development of terahertz communication technology in the future [31, 32].

1. Semiconductor technology, including RF, analog baseband and digital logic, *etc*.;
2. Research low-power high-speed baseband signal processing, corresponding low-complexity integrated circuit design, and high-speed communication baseband platform based on terahertz;
3. Terahertz modulation/demodulation technology, including direct modulation, mixing modulation, and photoelectric modulation, *etc* [17];
4. Waveform and channel coding;
5. Synchronization mechanism, for example, tracking and synchronization technologies of hundreds of orders of magnitude antenna elements in dense access scenarios;
6. Channel measurement and modeling of terahertz space and terrestrial communication [18].

The research on the above-mentioned several aspects of technical issues needs to be considered comprehensively in order to strike a balance between the performance, complexity and power consumption of terahertz communication.

In addition, the ITU has decided to assign 0.12 THz and 0.2 THz to the new generation of mobile communications in terms of spectrum supervision. However, the regulatory rules for the band exceeding 0.3 THz have not yet been unified globally. The ITU level and the WRC meeting need to work together and actively promote to reach a consensus.

SUMMARY

This chapter mainly presents the research status of 6G mobile networks with the focus on spectrum composition of next generation communication system. Since the 5G networks could hardly meet all the requirements from future developing society of 2030 and beyond, 6G research has become inevitable and attracted the interest all over the world. In order to achieve remarkably faster data rate than 5G, the higher frequency spectrums than previous generation's have to be exploited. We have specially highlighted the advantages of mmWave and Terahertz Spectrums with details. In this chapter, we also introduce the international Organizations for 6G Standardization and the progresses of 6G technology in different countries.

CONSENT FOR PUBLICATION

Not applicable.

CONFLICT OF INTEREST

The author declares no conflict of interest, financial or otherwise.

ACKNOWLEDGEMENTS

Declared none.

REFERENCES

[1]   S. Dang, "Osama Amin, BasemShihada and Mohamed-Slim Alouini, "What should 6G be?", *Nature Electronics,* vol. 3, pp. 20-29, 2020.
       [http://dx.doi.org/10.1038/s41928-019-0355-6]

[2]   Z. Zhang, and Y. Xiao, "6G wireless networks: vision, requirements, architecture, and key technologies", *IEEE Veh. Technol. Mag.,* vol. 14, no. 3, pp. 28-41, 2019.
       [http://dx.doi.org/10.1109/MVT.2019.2921208]

[3]   M. Katz, M. Matinmikko-Blue, and M. Latva-Aho, "6Genesis Flagship Program: Building the Bridges Towards 6G-Enabled Wireless Smart Society and Ecosystem", *2018 IEEE 10th Latin-American Conference on Communications (LATINCOM), Guadalajara, Mexico,* pp. 1-9, 2018.
       [http://dx.doi.org/10.1109/LATINCOM.2018.8613209]

[4]   I. Tomkos, D. Klonidis, E. Pikasis, and S. Theodoridis, "Toward the 6G Network Era: Opportunities and Challenges", *IT Prof.,* vol. 22, no. 1, pp. 34-38, 2020.
       [http://dx.doi.org/10.1109/MITP.2019.2963491]

[5]   C. Sergiou, M. Lestas, P. Antoniou, C. Liaskos, and A. Pitsillides, "A Prospective Look: Key Enabling Technologies, Applications and Open Research Topics in 6G Networks", *IEEE Access,* vol. 8, pp. 89007-89030, 2020.
       [http://dx.doi.org/10.1109/ACCESS.2020.2993527]

[6]   M. Giordani, M. Polese, M. Mezzavilla, S. Rangan, and M. Zorzi, "Toward 6G networks: Use cases and technologies", *IEEE Commun. Mag.,* vol. 58, no. 3, pp. 55-61, 2020.
       [http://dx.doi.org/10.1109/MCOM.001.1900411]

[7]   A. Yastrebova, R. Kirichek, Y. Koucheryavy, A. Borodin, and A. Koucheryavy, "Future networks 2030: architecture & requirements", *2018 10th International Congress on Ultra Modern Telecommunications and Control Systems and Workshops (ICUMT)* pp. 1-8, 2018.
       [http://dx.doi.org/10.1109/ICUMT.2018.8631208]

[8]   K. Samdanis, and T. Taleb, "The Road beyond 5G: A Vision and Insight of the Key Technologies", *IEEE Netw.,* vol. 34, no. 2, pp. 135-141, 2020.
       [http://dx.doi.org/10.1109/MNET.001.1900228]

[9]   S. Chen, Y-C. Liang, and S. Sun, "Vision, Requirements, and Technology Trend of 6G: How to Tackle the Challenges of System Coverage, Capacity, User Data-Rate and Movement Speed", *IEEE Wirel. Commun.,* vol. 27, no. 2, pp. 218-228, 2020.
       [http://dx.doi.org/10.1109/MWC.001.1900333]

[10]   P. Yang, Y. Xiao, M. Xiao, and S. Li, "6G Wireless Communications: Vision and Potential Techniques", *IEEE Netw.,* vol. 33, no. 4, pp. 70-75, 2019.
       [http://dx.doi.org/10.1109/MNET.2019.1800418]

[11]   X. You, and C-X. Wang, "Towards 6G wireless communication networks: Vision, enabling technologies, and new paradigm shifts", *Sci. China Inf. Sci.,* 2020.
       [http://dx.doi.org/10.1007/s11432-020-2955-6]

[12]   C.T. Neil, M. Shafi, and P.J. Smith, "Impact of microwave and mmWave channel models on 5G systems performance", *IEEE Trans. Antenn. Propag.,* vol. 65, no. 12, pp. 6505-6520, 2017.

[http://dx.doi.org/10.1109/TAP.2017.2759958]

[13] R. Adeogun, G. Berardinelli, P.E. Mogensen, I. Rodriguez, and M. Razzaghpour, "Towards 6G in-X Subnetworks with Sub-Millisecond Communication Cycles and Extreme Reliability", *IEEE Access,* vol. 8, pp. 110172-110188, 2020.
[http://dx.doi.org/10.1109/ACCESS.2020.3001625]

[14] Y. Henri, "ITU world radio communication conference (WRC-15) allocates spectrum for future innovation", *Air Space Law,* vol. 41, no. 2, pp. 119-128, 2016.

[15] M.M. Sohul, M. Yao, T. Yang, and J.H. Reed, "Spectrum access system for the citizen broadband radio service", *IEEE Commun. Mag.,* vol. 53, no. 7, pp. 18-25, 2015.
[http://dx.doi.org/10.1109/MCOM.2015.7158261]

[16] M. Matinmikko-Blue, S. Yrjölä, and P. Ahokangas, *Spectrum Management in the 6G Era: The Role of Regulation and Spectrum Sharing*, 2020.

[17] G. Naik, J. Liu, and J-M. Park, "Coexistence of wireless technologies in the 5 GHz bands: A survey of existing solutions and a roadmap for future research", *IEEE Comm. Surv. and Tutor.,* vol. 20, no. 3, pp. 1777-1798, 2018.
[http://dx.doi.org/10.1109/COMST.2018.2815585]

[18] K. Zheng, L. Zhao, J. Mei, M. Dohler, W. Xiang, and Y. Peng, "10 Gb/s HetSNets with Millimeter-Wave Communications: Access and Networking – Challenges and Protocols", *IEEE Commun. Mag.,* vol. 53, no. 1, pp. 222-231, 2015.
[http://dx.doi.org/10.1109/MCOM.2015.7010538]

[19] L. Zhu, Z. Xiao, X. Xia, and D. Oliver Wu, "Millimeter-Wave Communications with Non-Orthogonal Multiple Access for B5G/6G", *IEEE Access,* vol. 7, pp. 116123-116132, 2019.
[http://dx.doi.org/10.1109/ACCESS.2019.2935169]

[20] S. Rangan, and S. Theodore, "Rappaport and E. Erkip, "Millimeter-Wave Cellular Wireless Networks: Potentials and Challenges", *Proc. IEEE,* vol. 102, no. 3, pp. 366-385, 2014.
[http://dx.doi.org/10.1109/JPROC.2014.2299397]

[21] O. Kanhere, and T.S. Rappaport, "Position Locationing for Millimeter Wave Systems", *IEEE Global Communications Conference (GLOBECOM),* 2018pp. 206-212 Abu Dhabi, United Arab Emirates

[22] S.A. Busari, K. Mohammed, and S. Huq, "Millimeter-Wave Massive MIMO Communication for Future Wireless Systems: A Survey", *IEEE Comm. Surv. and Tutor.,* vol. 20, no. 2, pp. 836-879, 2018.
[http://dx.doi.org/10.1109/COMST.2017.2787460]

[23] T.S. Rappaport, and Y. Xing, "Wireless Communications and Applications Above 100 GHz: Opportunities and Challenges for 6G and Beyond", *IEEE Access,* vol. 7, pp. 78729-78757, 2019.
[http://dx.doi.org/10.1109/ACCESS.2019.2921522]

[24] Int; Telecommun; Union (ITU), "Technology trends of active services in the frequency range 275-3 000 GHz", *ITU-Recommendation SM.,* pp. 2352-0, 2015.

[25] H. Viswanathan, and P.E. Mogensen, "Communications in the 6G Era", *IEEE Access,* vol. 8, pp. 57063-57074, 2020.
[http://dx.doi.org/10.1109/ACCESS.2020.2981745]

[26] K. Tekbıyık, A.R. Ekti, G.K. Kurt, and A. Görçinad, "Terahertz band communication systems: Challenges, novelties and standardization efforts", *Phys. Commun.,* vol. 35, no. 100700, 2019.
[http://dx.doi.org/10.1016/j.phycom.2019.04.014]

[27] P. Planinšič, B. Pongrac, and D. Gleich, "THz communications and relations to frequency domain THz spectroscopy", *15th International Conference on Telecommunications (ConTEL),* Graz, Austria, pp. 1-5, 2019.
[http://dx.doi.org/10.1109/ConTEL.2019.8848561]

[28] M. Sakai, and K. Kamohara, "Experimental Field Trials on MU-MIMO Transmissions for High SHF

Wide-Band Massive MIMO in 5G", *IEEE Trans. Wirel. Commun.,* vol. 19, no. 4, pp. 2196-2207, 2020.
[http://dx.doi.org/10.1109/TWC.2019.2962766]

[29]   I.F. Akyildiz, J.M. Jornet, and C. Han, "Terahertz band: Next frontier for wireless communications", *Phys. Commun.,* vol. 12, no. 4, pp. 16-32, 2014.
[http://dx.doi.org/10.1016/j.phycom.2014.01.006]

[30]   K. Mohammed, S. Huq, and S.A. Busari, "Terahertz-Enabled Wireless System for Beyond-5G Ultra-Fast Networks: A Brief Survey", *IEEE Netw.,* vol. 33, no. 4, pp. 89-96, 2019.
[http://dx.doi.org/10.1109/MNET.2019.1800430]

[31]   H. Elayan, O. Amin, B. Shihada, R.M. Shubair, and M. Alouini, "Terahertz Band: The Last Piece of RF Spectrum Puzzle for Communication Systems", *IEEE Open Journal of the Communications Society,* vol. 1, pp. 1-32, 2020.
[http://dx.doi.org/10.1109/OJCOMS.2019.2953633]

[32]   Z. Chen, and X. Ma, "A survey on terahertz communications", *China Commun.,* vol. 16, no. 2, pp. 1-35, 2019.
[http://dx.doi.org/10.23919/JCC.2019.09.001]

CHAPTER 2

Millimeter Wave Communication Technology

Aart W. Kleyn[1], Wei Luo[2,*] and Bo Yin[2]

[1] *Center of Interface Dynamics for Sustainability, Institute of Materials, China Academy of Engineering Physics, Chengdu 610200, Sichuan, China*

[2] *College of Electronic Engineering, Chongqing University of Posts and Telecommunications, Chongqing, China*

Abstract: Millimeter-wave plays an indispensable role in the new generation of mobile communication because of its abundance of unexplored resources, which can be used to meet the requirements of greater bandwidth and ultra-high data rate. This chapter firstly introduces the development, characteristics, and applications of millimeter-wave. Then the application of millimeter-wave in mobile communication systems is described in detail, and the common channel model is listed. Secondly, a large-scale MIMO system, precoding technology, and three kinds of beam forming structures are introduced. Finally, combined with the current development of mobile communication, the requirements on the millimeter-wave devices for the new generation of mobile communication system are summarized, and some typical millimeter-wave devices are listed.

Keywords: Backhaul, Channel model, Millimeter wave, Propagation model, Wireless communication, Wireless network.

INTRODUCTION

The birth of each generation of mobile communication technology has provided great convenience for our life. The emergence of second-generation mobile communication technology (2G) has provided us with convenient voice calls. In the 3G era, we can use the mobile Internet to achieve some basic network functions. For the last ten years, high-speed mobile Internet in the age of 4G has brought video calling and mobile movie-watching into our lives. The 5G is the era of the Internet of everything, which takes the advantages of higher channel capacity, faster network speeds, and lower latency compared to 4G. Based on these characteristics, telemedicine, smart home, artificial intelligence (AI), the Internet of things (IoT), and so on will be developed and applied better.

* **Corresponding author Wei Luo:** College of Electronic Engineering, Chongqing University of Posts and Telecommunications, Chongqing, China; Tel: 0086 23 62471418; Fax: 0086 23 62471716; E-mail: luowei1@cqupt.edu.cn

5G spectrum resources are divided into FR1 and FR2 by frequency. FR1 is the low-frequency part of 450~6000 MHz, which is also named Sub 6 GHz. FR2 is part of the millimeter wave (mmWave) frequency band (24.25~52.6 GHz). The specific division of the 5G spectrum in China was completed in 2018. Since low-frequency electromagnetic wave has a long wavelength, it has the small path loss and the large coverage area. Meanwhile, the relevant communication technology of low-frequency band is mature, and the application cost is low, which can quickly realize 5G deployment. Therefore, Sub 6 GHz is the main frequency band for the development of China's mobile communication at this stage. However, the mmWave frequency band is still an important frequency band for the future development of 5G in China. Compared with Sub 6 GHz, the bandwidth of the mmWave frequency band is GHz, which can realize a higher transmission rate and accommodate more data information. It is often located in a densely populated area and has high application value in the industrial field.

In recent years, with the development of mobile communications, the research on mmWave communications has become a major international focus. The mmWave communication and massive MIMO are considered to be the two key technologies of 5G. Because of the large bandwidth of the mmWave frequency band, the high transmission rate requirements of mobile communication systems for the new generation could be met. Since the mmWave has a short wavelength, the size of large antenna arrays could be miniaturized. Meanwhile, the use of the directional beamforming gain provided by the massive MIMO system can effectively compensate for the path loss of mmWave signals in the wireless channel. The current research on the new generation of mobile communications, from the planning and authorization of spectrum resources with specific needs to the research and development of various mmWave equipment and the construction of a new generation of mobile communication systems, has made tremendous progress. In general, the challenges faced by millimeter-wave frequency communications are mainly considered as high-frequency devices. Related high-frequency core components mainly include power amplifiers, low-noise amplifiers, phase-locked loop circuits, filters, high-speed and high-precision digital-to-analog and analog-to-digital converters, array antennas, *etc*.

CHARACTERISTIC AND APPLICATION OF MILLIMETER WAVE

With the acceleration of the informatization of human society, the application scope of the electromagnetic spectrum is expanded. In the 1970s, the World Radio Conference held by the International Telecommunication Union (ITU) divided and allocated the 30~70GHz frequency band used by communication services. In recent years, due to the continuous growth of demand for broadband and large-capacity information transmission, personal communications, and military confidentiality/anti-jamming communications, the mmWave and even submillimeter-wave fields have become an extremely active field in the research, development, and utilization of international electromagnetic spectrum resources, which contains abundant information resources. To increase the communication capacity and avoid channel congestion and mutual interference, the communication frequency must be developed into a higher frequency band. Therefore, the developments of electromagnetic spectrum resources in the mmWave and terahertz bands are recently the key research fields in electronic science, and the spectrum resource of mmWave band has significant application values. Generally, electromagnetic waves with a frequency range of 30~300GHz are called mmWave, and can be divided into several frequency bands. The frequency band codes and frequency ranges are shown in Table **1**.

Table 1. Millimeter wave frequency division.

Band Code	K	K_a	Q	U	E	W	F	G	M
Freq (GHz)	18~26.5	26.5~40	33~50	40~60	60~90	75~110	90~140	140~220	170~260

Characteristic of Millimeter Wave

The wavelength of millimeter wave is located in the overlap range of microwave and infrared waves, which has the characteristics of both ones. The theory and technology of millimeter wave are the extension of microwave to high frequency and the development of light waves to low frequency [1].

Usable Frequency Bandwidth and Large Information Capacity

The mmWave frequency band ranges from 30 to 300 GHz, with a bandwidth of up to 270GHz, which is more than 10 times the full bandwidth from DC to microwave. Considering the atmospheric absorption, there are mainly four propagating windows for mmWave. The total bandwidth of these four windows can reach 135GHz, which is five times the sum of the bandwidth of each band below the microwave. Therefore, the main advantages of mmWave communication are bandwidth and large information capacity, which can be used for multi-channel communication and television image transmission. Furthermore, the high transmission rate is conducive to the realization of communication with a low probability of interception, such as spread spectrum communication and frequency hopping communication. High-loss frequencies (such as 60 GHz, 120 GHz, and 180 GHz) can also be used for military confidential communications and satellite communications.

Short Wavelength in Millimeter Wave Band

Compared with centimeter-wave, millimeter-wave has a shorter wavelength. Therefore, its quasi-optical characteristics can be used to achieve higher angular resolution capability in the case of relatively small antenna aperture. Moreover, higher positioning accuracy in space could be achieved for conducting target navigation, positioning and tracking. The ratio of target signal to passive jamming one is higher, and it has better anti-jamming stability to active jamming.

Significant Impact of the Atmosphere on Millimeter Wave Transmission

When electromagnetic waves propagate in the earth's atmosphere, the atmosphere absorbs and emits rays of various wavelengths. While this effect is small at low frequency, the atmosphere absorption is prominent for the mmWave frequency band and varies with frequency. Atmospheric attenuation is severe in some specific frequency bands. When electromagnetic waves of a certain frequency are incident on atmospheric molecules, the energy is transferred due to resonance. If the molecules return to their original state, the energy is released at the same frequency. The released energy is radiated randomly, and the results are the absorption and attenuation of electromagnetic waves and noise contribution. Various atmospheric gases have resonance phenomena in the mmWave frequency band, but only oxygen and water vapor have obvious absorption characteristics. Oxygen has a magnetic dipole moment. There are more than 40 resonance lines around 60GHz and one resonance line at 118.75GHz. Water vapor has an electric dipole moment, of which the resonance spectrum is about 22 GHz and 183GHz in mmWave band. The frequency range of the transmission line and the atmospheric attenuation are shown in Table **2**.

Table 2. Transmission band and its atmospheric attenuation.

Transmission band(GHz)	24~48	70~110	120~150	190~300
Atmospheric attenuation(dB/km)	0.1~0.3	0.3~1	1~2.5	2~10

Compared with light waves, mmWave propagates with atmospheric windows (When mmWave and submillimeter wave propagate in the atmosphere, some attenuation caused by the resonance absorption of gas molecules is a minimum value). The influence of heat radiation source is little. The correct selection of frequencies is crucial for the design of mmWave systems.

Application of Millimeter Wave

Compared with microwaves, mmWave corresponds to shorter wavelengths, and has a certain overlap with the low frequency end of terahertz waves [2]. In view of the various spectrum and propagation characteristics, the application range of mmWave is extensive.

Millimeter Wave Communication System

With the continuous advancement of global informatization, people's requirements for the transmission rate, transmission bandwidth, and transmission quality of communication systems continue to increase, which lead to the birth of millimeter-wave wireless broadband access technology. In addition, the satellite communication system has launched nearly 200 communication satellites into synchronous orbits since the 1990s. The communication frequency band is mainly occupied by the C-band, followed by the Ku-band, causing the problem of congested satellite communication spectrum. The introduction of mmWave can increase the capacity and transmission rate of satellite communications. In addition, the mmWave has strong anti-interference ability and concealment, which makes satellite communications more secure and reliable.

Millimeter Wave Weapon

Compared with infrared guided or laser guided weapons, millimeter-wave guided weapons have a stronger ability to penetrate smoke, dust, or clouds, and have lower attenuation and atmospheric loss in the transmission window. Thus, they can be used in very harsh environments, such as severe weather or a complex battlefield, and operate normally to achieve the effect of precision strikes. The other weapon application is directly use mmWave to generate high-power microwave pulse signals to permanently damage electronic equipment. For the human body, since the skin depth of mmWave is only 0.4mm, this high-power microwave pulse signal will not endanger lives, but will cause an unbearable burning sensation on the epidermis and loss of combat effectiveness.

Millimeter Wave Imaging Technology

Traditional X-ray imaging technology is used in most airport and station security inspection occasions, but the radiation generated by this equipment is harmful to the human body. Thus, multiple security inspections are not recommended in a short time, otherwise, it may cause irreversible damage to the human body. The mmWave can penetrate common shielding materials such as clothing, plastic or paper, and has a high resolution. It can accurately identify dangerous goods carried without generating ionizing radiation. Therefore, compared with traditional X-ray security inspection technology, it is more suitable for the human body safety inspection field.

Millimeter Wave Radar

Radar technology is also an important part of national defense strength. The rapid development of high-performance mmWave sources has greatly improved the performance of radar [3]. For example, it can distinguish smaller targets and distinguish multiple targets at the same time, and it can adapt to complex weather environment, *etc*. Simultaneously, mmWave radar can be smaller in size and lighter in weight, which meets the needs of the development of military electronic equipment.

MILLIMETER WAVE TECHNOLOGY IN MOBILE COMMUNICATION

As a subdivision of microwave, mmWave shows many common characteristics in communication transmission. mmWave has been diffusely used in various fields because of good anti-jamming ability, good security characteristic, wide bandwidth, low power consumption and so on. The mmWave communication technology includes two aspects: further development of light wave to low frequency, extension of microwave to high frequency. At present, the frequency band resources used in the world are very scarce, and the spectrum of the wireless telecommunications system used is lower than that of 6 GHz. However, the spectrum resources in the 3~60 GHz range are sufficient and have not been effectively developed and used. For this reason, the new generation mobile communication takes 9.9~86GHz band as the key interval and makes a more in-depth exploration of Internet wide area coverage based on LTE (Long Term Evolution) technology [4]. This band is used to create a super-dense network, and the LTE layer and UDN (Ultra-Dense Network) layer will be covered double. With the increase of the number of base station antennas, a large number of antennas can be arranged in insufficient space, and the wavelength of communication is not too long.

With the increase in the number of wireless devices and the scarcity of spectrum resources, the limited bandwidth can only be shared on the narrow spectrum, which greatly affects the user experience. 5G communication uses mmWave (26.5~300 GHz) technology, which increases the rate by increasing the bandwidth. The available bandwidth for 28 GHz is about 1 GHz, while the available bandwidth for 60 GHz is 2 GHz.The mmWave has outstanding advantages in improving bandwidth and signal quality and covering.

At present, through large-scale multi-input and output technology in the mmWave band, Bell Laboratories in China has achieved the goal of increasing capacity and communication efficiency. Using prototypes with peak transport speeds higher than 50 Gbps, the lab has increased the spectrum speed to 100 bps per 28 GHz at the mmWave band 28 GHz, enabling users to download faster by using the Internet and deliver as much as 100 megabytes of information in a very short period of time.

To sum up, mmWave communication occupies an important application field with its own advantages, and cannot be replaced in the short term. Based on the characteristic of mmWave, the very high-speed communication of the new generation mobile communication could be supported by the mmWave technology. The application of large-scale antenna technology to mmWave band communication to achieve a new generation of mobile communication system is a vision generally accepted by the industry, which can effectively support the bandwidth and spectrum efficiency requirements of the new generation mobile communication system.

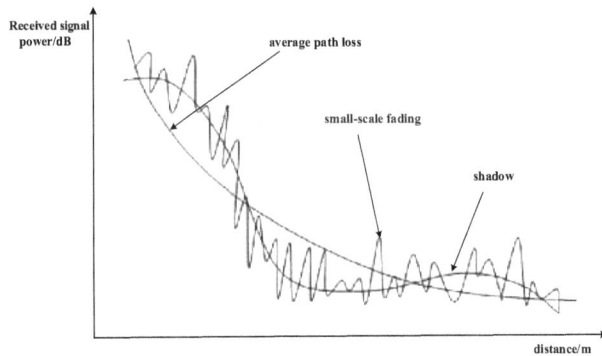

Fig. (1). Schematic diagram of large-scale fading and small-scale fading.

Millimeter Wave Propagation Model and Channel Model

As the transmission medium in the process of wireless communication, the propagation characteristic of the electromagnetic wave in space is the basis of transmission technology configuration, antenna structure design and the wireless network plan in wireless communication system. For the mmWave band, because the target size and wavelength in the propagation environment are comparable, the propagation in space is different from the low frequency propagation law, and is closely related to the propagation environment. The mmWave attenuates severely in the atmospheric propagation, and the propagation distance is seriously suppressed. In addition, the electromagnetic wave attenuation caused by the absorption of oxygen molecules, water vapor molecules and rainfall factors are also closely related to the frequency [5]. Electromagnetic waves will show different propagation mechanisms such as reflection, diffraction, scattering or transmission. mmWave propagation models are mainly divided into two categories: large-scale propagation model and small-scale propagation model, as shown in Fig. (**1**).

The purpose of the large-scale propagation model is to describe the variation of the received signal power with the transceiver distance over a long distance between the transmitter and the receiver.

Large-scale Loss Propagation Model

The large-scale loss propagation model is a description of the change characteristics of the signal strength of electromagnetic waves transmitted over long distances in a wireless channel. In a simple wireless channel environment, the large-scale loss characteristic describes the proportional relationship between receiver signal strength and distance. For several known large-scale loss models, different wireless channel environments have different path loss models. The path loss reflects the energy reduction of the electromagnetic wave after passing through the wireless channel. It is generally expressed in the amount of power reduction and reflects the change in the receiver's received power in the effective long-distance transmission range. In a general channel environment, when the transmitted power is constant, the path loss is proportional to the transmission distance. As the propagation distance increases, the signal strength is constantly decreasing. Therefore, a variable defined by the decreasing trend is called the path loss index. The larger the loss index, the faster the signal strength decreases as the distance increases. The lowest value in the communication system is not lower than the receiver's receiving threshold. Path loss index is closely related to electromagnetic wave frequency, antenna properties and channel environment.

There are various large-scale path loss models (deterministic, empirical, and random), and all of them provide true path loss models for wireless channels based on measured parameters. The model is divided into single-frequency and multi-frequency path loss models, including the following typical large-scale path loss models.

The most widely used single-frequency large-scale path loss models are the CI (Close-In free space reference distance path loss model) and FI (Floating-Intercept path loss model). The expression of CI model is as follows:

$$PL^{CI}(f,d)_{[dB]} = FSPL(f,d_0) + 10n \log_{10}\left(\frac{d}{d_0}\right) + X_\delta^{CI} \qquad (1)$$

$$FSPL(f,d_0) = 10 \log_{10}(\frac{4\pi d_0}{\lambda})^2 \qquad (2)$$

where X_δ^{CI} is a zero-mean Gaussian random variable based on the standard deviation δ, $FSPL(f, d_o)$ represents the path loss when the frequency is constant and the reference distance d_o, and n is PLE (Path Loss Exponent).

Generally, the value of d_o is in the far field, and d should be much larger than the d_o. In the case of outdoor low frequency, d_o is generally 0.1 ~ 100km, but for high frequency indoor environment, d_o is generally 1 m. The CI model is only determined by PLE, and the least square method is usually used to fit the data. SF (Shadow Fading) can be expressed by the standard deviation of the measured data, defined as follows:

$$SF(dB) = X_\delta(0, \delta) \tag{3}$$

$$\delta(dB) = \sqrt{\frac{1}{N-1} \sum_{i,j}^{i,j=1} \left[PL(d_{i,j}) - \overline{PL(d_{i,j})} \right]^2} \tag{4}$$

where N represents the number of paths between the transmitter and receiver, and δ represents the standard deviation of the measured data.

In the process of outdoor wireless propagation, electromagnetic waves are blocked by buildings and topography, and a semi-blind area is formed behind the shielding objects. The propagation mechanism of electromagnetic waves will change due to the change of these uncertain factors. In the process of indoor wireless transmission, the main occlusion comes from walls, people, *etc*. Therefore, shadow fading will directly affect the wireless communication network coverage and communication quality.

Small-scale Propagation Model

The purpose of the small-scale propagation model is to characterize the small-scale fading characteristics of electromagnetic waves with rapid changes in signal intensity in a small range. Small-scale fading refers to the rapid change of the received signal in a short time or short distance, and small-scale fading is also called fast fading. The transmitted signal propagates through multiple paths and reach the receiver based on different propagation mechanisms. At this time, the received signal is a set of signals with different amplitudes and phases transmitted with different paths, and these signals will interfere with each other, resulting in small-scale fading. The signal propagated through multiple paths is called a multipath signal, and the multipath signal has different signal strength and signal bandwidth in the process of propagation. For mmWave frequency band, on the one hand, the frequency selectivity of mmWave wireless channel is more serious than that of low frequency, and there are differences in small-scale fading characteristics between different frequency bands. On the other hand, the penetration ability of mmWave signal is weak, and the transmitted signal can be received in the NLOS (Non Line of Sight) environment. In order to achieve the coverage of the NLOS area, it is necessary to study the small-scale angular dispersion characteristics of the mmWave signal.

In the mmWave channel model, the geometric space-based channel model usually uses CIR (Channel Impulse Response) to reflect the small-scale characteristics of multipath channels, and in the measurement of radio propagation, small-scale propagation parameters are often extracted based on the CIR of multipath signals. The impulse response model of the mmWave multipath channel can be represented as a function of time and space angle

$$h(t, \vec{\Theta}, \vec{\Phi}) = \sum_{K=1}^{K} \alpha_K e^{j\varphi_K} \delta(t - \tau_K) \cdot \delta(\vec{\Theta} - \vec{\Theta}_K) \cdot \delta(\vec{\Phi} - \vec{\Phi}_K) \qquad (5)$$

where K is the number of multipath, t represents time, α_K, φ_K and τ_K represent the amplitude, phase and relative propagation delay of multipath K, respectively, $\vec{\Theta}_K$ and $\vec{\Phi}$ denote the angle of departure and the angle of arrival, respectively, both of which contain the horizontal angle and pitch angle information. $\vec{\Theta}_K$ represents the vector (θ_{rx}, ψ_{rx}) composed of AoD (Azimuth angle Of Departure) and EoD (Elevation angle Of Departure) of multipath K, and Φ represents the vector (θ_{rx}, ψ_{rx}) composed of AoA (Azimuth angle Of Arrival) and EoA (Elevation angle of Arrival) of multipath K.

Millimeter Wave Channel Model

Channel model is the basis of wireless communication system design and communication algorithm technology evaluation. Thus, accurate channel modeling is very important for wireless communication. In order to simulate the real signal transmission process and summarize the channel characteristics, it is necessary to carry out a large number of measurements on typical scenes to obtain effective data support and make the results more reliable. At present, most of the mmWave channel modeling methods are time cluster and spatial lobe method, as well as channel model method based on ray tracing technology. Some remarkable characteristics of mmWave are introduced as an important reference in the modeling process [6]. For example, the mmWave represented by 60 GHz is just on the absorption peak of oxygen and water vapor, and the electromagnetic wave signal propagates in the atmosphere for a long distance. Some studies have shown that the attenuation of 60 GHz signal on some materials is much higher than that of low-frequency signal, and the attenuation also varies with the materials. In addition to the above characteristics, there are two more important characteristics of mmWave. Firstly, the mmWave is easier to be blocked by the human body and produce significant attenuation. Meanwhile, the mmWave channel is non-stationary, due to the relatively frequent and complex human activities. Secondly, based on a large number of measured data, it is found that the mmWave presents significant characteristics cluster arrival phenomenon. Thus, many standards on the mmWave channel model take this feature as the basis of modeling. The flow chart of design solutions for mmWave channel modeling is shown in Fig. (2).

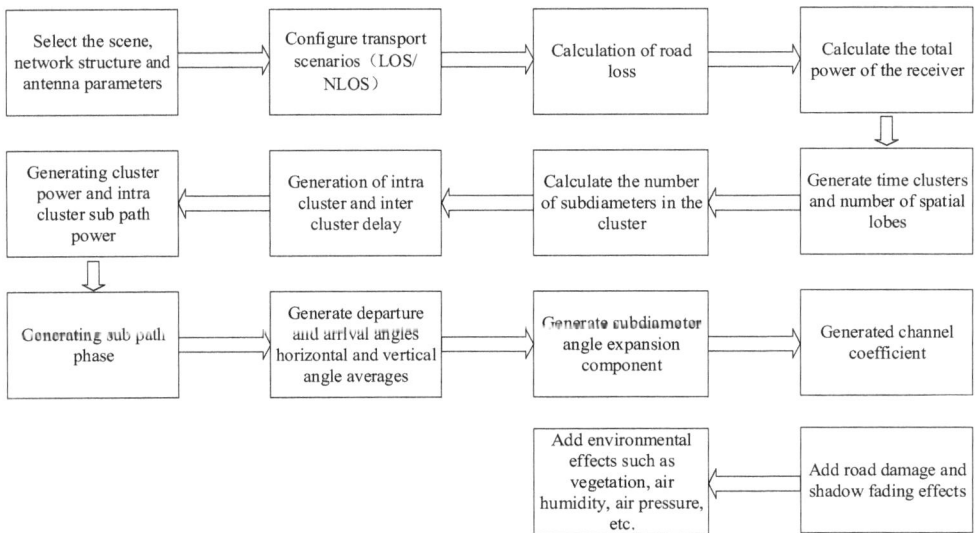

Fig. (2). Channel modeling flow chart.

At present, there are three international standards for the mmWave channel model, which are IEEE802.15.3c, IEEE802.11ad and ECMA-387. And the high-frequency channel models that have been completed and in progress mainly include the channel models proposed by research institutions, various research project groups and standardization organizations, as well as those proposed by 3GPP.

The WiGig Alliance is committed to the study of 60 GHz WLAN systems and launched the corresponding IEEE802.11ad channel model in 2009. The model supports three scenarios: conference room, cubicle and living room. The path loss and small-scale parameters of each scene at LOS (Line of Sight) / NLOS are given in detail. In this model, according to the indoor propagation characteristics of 60 GHz electromagnetic waves, the concept of the cluster is introduced, the human blocking model is established, and the blocking probability of the cluster is set. The unblocked cluster needs to generate parameters such as time, amplitude, phase and polarization characteristics of the sub-diameter in the cluster. In addition, the modified model provides three kinds of antenna models: omni-directional antenna, directional controllable antenna and phased array antenna.

MiWEBA Channel Model

The MiWEBA research project is devoted to the research and development of mmWave heterogeneous networks, which connects mmWave UWB base stations and cellular networks through backhaul/forward. The project aims to expand the network capacity by 1000 times and solve the problem of capacity limitation at a reasonable cost. The fifth working group of the project is responsible for the measurement and research of the mmWave channel model, and develops its own mmWave band channel model. The MiWEBA channel model is based on the 3GPP 3D channel model framework. The measurement data and RT (Ray Tracing,) statistical methods are used to develop the channel model, with emphasis on testing the path loss model in the 57 GHz ~66 GHz band. As the limited measurement activities cannot effectively solve the problems such as vegetation loss and human shadow fading caused by mobile vehicles, the MiWEBA project developed a new accurate (Quasi-Deterministic, Q-D) method to model the outdoor and indoor channels of 60 GHz. The mmWave channel impulse response of the model consists of several deterministic strong rays (D rays) and many relatively weak random rays (R rays). The structure of the channel model can flexibly describe the geometric characteristics of the scene: reflection attenuation and scattering, ray blocking and moving effects. Three main scenarios are defined in the MiWEBA project proposal: open areas (university campuses), streets and hotel lobbies. The model has good versatility, and the appropriate link model parameters are selected on the basis of experimental measurement and ray tracing modeling according to the requirements. For example, changing the Tx antenna parameters can be extended to the D2D (Device to Device) link.

METIS Channel Model

Compared with other project teams, the METIS (Mobile and wireless communications Enablers for the Twenty-twenty Information Society) project develops new channel models based on a wide range of measurement activities and analysis, supplemented by computer simulations. METIS channel models include map-based models, random models or a combination of the two models, which provide a flexible and scalable channel model framework to meet a variety of simulation requirements in terms of accuracy and computational complexity. Map model is a ray tracing model based on simplified map, which includes important propagation mechanisms such as diffraction, specular reflection, diffuse reflection, blocking and so on. Therefore, in view of the fact that the model provides accurate and real spatial channel characteristics, it is suitable for large-scale MIMO and D2D path loss model. The stochastic model is based on the geometric random channel model evolved from the WINNER/3GPP channel model. The statistical parameters provided have mmWave characteristics, and the path loss model is frequency dependent. The hybrid model has a flexible and scalable model framework, which can effectively balance the complexity and accuracy of simulation. For example, shadow fading can be based on map model, while small-scale fading can be based on random model.

New York University Channel Model

The team of Professor Rappaport of New York University first announced the outdoor two-dimensional mmWave (28 GHz) channel model in 2014. The model is based on the 3GPP channel model framework and uses measurement data and ray tracing statistical methods to develop the channel model. The test scenarios are mainly in the downtown, campus, indoor and outdoor under the dense city (New York City Manhattan area). Subsequently, the team proposed a three-dimensional channel model based on random geometry. The large-scale and small-scale parameters in the model meet different random distributions (similar to 3GPP). The model defines two small-scale parameters, time-domain clusters and spatial lobes. The number of clusters and sub-paths in each scene is uniformly distributed. The number of spatial lobes obeys Poisson distribution, and the remaining channel parameters are similar to 3GPP. In addition, the model provides three path loss models for beam combining, directional and omnidirectional. The channel model proposed by this team is only for the frequency bands it measures (28 GHz, 38 GHz, 60 GHz, 73.5 GHz) and does not adapt to all frequency spectrums from 6 GHz ~100 GHz.

mmMAGIC Channel Model

The mmMAGIC (Millimetre Wave Based Mobile Radio Access Network for Fifth Generation Integrated Communications) has developed a high-frequency channel model for the frequency range of 6 GHz ~100 GHz, and carried out 20 kinds of propagation environments (urban blocks, open squares, indoor offices, airports and indoor to outdoor) in eight frequency bands between 6 and 100 GHz. The mmMAGIC channel model was developed in parallel with 3GPP and ITU-R, and some development methods were adopted by the latter. The model is based on a geometric random model and consists of a basic model and an additional model. Among them, the functions of the additional model include: ground reflection and blocking effects, large bandwidth and large antenna arrays, spatial consistency and building penetration loss modeling. The measurement scheme of mmMAGIC is similar to that of METIS. Because the measurement data is statistically limited, supplementary channel data is generated through ray tracing simulation, and then the measurement results are calibrated, so that additional parameters for random modeling can be derived.

5GCM Channel Model

5GCM (5G Channel Model) is a 5G mmWave channel model alliance initiated by the NIST (National Institute of Standards and Technology), including AT&T, China Mobile, Ericsson, Huawei, Intel, Nokia, Qualcomm, Samsung, New York University, University of Southern California and other research institutions and universities. This model is based on the three-dimensional channel model in 3GPP TR 36.873, using a combination of multi-frequency channel measurement and ray tracing simulation. Typical scenarios include UMa (Urban Macro), UMi (Urban Micro), InH (Indoor Hotspot) and O2I (Outdoor to Indoor). In addition to the path loss model, the model has many methods and parameter values consistent with the 3GPP TR 38.901 standard. For O2I penetration loss, for exterior wall buildings with different attenuation, two models are given, low loss and high loss.

This model provides important reference suggestions for the realization of spatial consistency. As the geometric position changes, the channel realization of large-scale parameters and small-scale parameters must be continuous. Firstly, the path loss including shadow fading should change smoothly as the UT moves. Secondly, small-scale parameters such as angle, power, and delay should also change dynamically with the position, which is of great significance for evaluating the mobility and beam tracking of 5G communications.

3GPP High Frequency Channel Model

In 2016, the 3GPP Standards Organization released the 3GPPTR38.900 high-frequency channel model, which is an extended version of the 3GPPTR36.873 three-dimensional model in the frequency range of 6cm 100GHz. It supports scenarios such as UMi, UMa, InH, and adds new channel characteristics in 5G scenarios, including spatial consistency, multicarrier aggregation, oxygen loss, ground reflection, large bandwidth and large antenna array. This model inherits the 5GCM blocking modeling method, further refines the polar coordinate expression, and divides the blocking objects into two types. In addition, two simplified models for link simulation are proposed: CDL and TDL. They are more flexible than the simplified model in WINNERII / SCME, and can set delay expansion or angle expansion according to the simulation requirements. Finally, the map hybrid model in METIS is used for reference. In 2017, the 3GPP Standards Organization released an enhanced version of the standard

3GPPTR38.901, which combines low-frequency and high-frequency channel models and extends the frequency range to 0.5~100GHz.

Shortage of Existing Millimeter Wave Channel and Its Future Development Trend

All the above channel models are based on high frequency band, which can meet the transmission characteristics of mm-wave large-scale MIMO system. In 5G and previous communication systems, the standardized channel model tends to be a general channel model framework based on geometric structure, which uses different parameters to describe different scenarios. At present, all the mm-wave standardized channel models are only for terrestrial communication networks, and need to be further expanded to meet the application requirements of 6G full coverage. In addition, driven by new application requirements and technical requirements, 6G mm-wave channel needs to introduce new performance metrics, such as higher spectral efficiency / energy efficiency / cost efficiency, higher transmission rate, lower delay, greater connection density, coverage, intelligent degree, security, *etc*.

In order to meet the above requirements and performance metrics of future mobile communication applications, 6G wireless communication network will adopt a new paradigm shifts and rely on new enabling technologies. The new paradigm shifts can be summarized as four trends: global coverage, all spectrum, full application and strong security. In order to provide global coverage, 6G wireless communication network will expand from land mobile communication to space, air, ground and sea integrated communication network, including satellite communication network, UAV, terrestrial ultra-dense networks, underground communication, marine communication and underwater acoustic communication network. In order to meet the application requirements of ultra-high transmission rate and ultra-high connection density, all spectrum including Sub-6 GHz, mm-wave, terahertz and optical band will be fully explored. Artificial intelligence (AI) and big data will be integrated with wireless communication network to better manage the communication network. Moreover, AI can better dynamically arrange network, caching and computing resources to improve system performance. Finally, in the design of the network, security factors should be taken into account, which is called built-in network security, including physical layer security and network layer security. Strong security is reflected in the design of the network at the same time, taking security factors into account, also known as network endogenous security, including physical layer security and network layer security. Therefore, deriving a new 6G general channel model framework will be more challenging. Due to the heterogeneous characteristics of 6G wireless channel and the different scale performance of radio waves, how to use 6G general channel model framework to describe 6G wireless channel accurately becomes a difficult and significant problem.

The Waveform of Millimeter Wave in Mobile Communication

With the advent of the new generation mobile communication era, the rapid growth of user data makes people consider using the carrier band above 6 GHz to meet the requirements of high bandwidth. The high frequency band focuses on the submillimeter wave band and mmWave band between 6~100GHz, which aims to provide larger communication capacity and lower delay expansion. The design of the mmWave band carrier waveform is still faced with severe challenges. The path loss and Doppler effect in the high frequency carrier increase obviously, and the hardware equipment of the signal processing system will have a nonlinear effect on the high frequency carrier. The high frequency effect of hardware devices and signal bandwidth and other related factors need to be fully considered in the design of receiving equipment and transmitting equipment, at the same time, due to the substantial increase of carrier frequency. There is no doubt that large-scale array antennas can better meet the requirements of high-frequency spectral efficiency [7].

Performance of Candidate Waves in High Frequency Band of Mobile Communication

MmMAGIC is a project launched in Europe to promote the establishment of a new generation of mobile communication mmWave wireless access network. The project has defined the relevant requirements and standards for new air ports above 5G 6GHz. In this plan, carrier candidate waveforms are divided into multi-carrier waveforms and single-carrier waveforms, and relevant indicators to measure waveform efficiency are defined, such as spectrum efficiency, complexity of hardware equipment, dynamic configuration of carrier parameters, power efficiency, out-of-band leakage and so on.

In mmMAGIC, the communication carriers above 6 GHz are mainly divided into Multi-Carrier (MC) and Single-Carrier (SC). The specific classification relationship is shown in Fig. (**3**).

Multi-Carrier | Single-Carrier

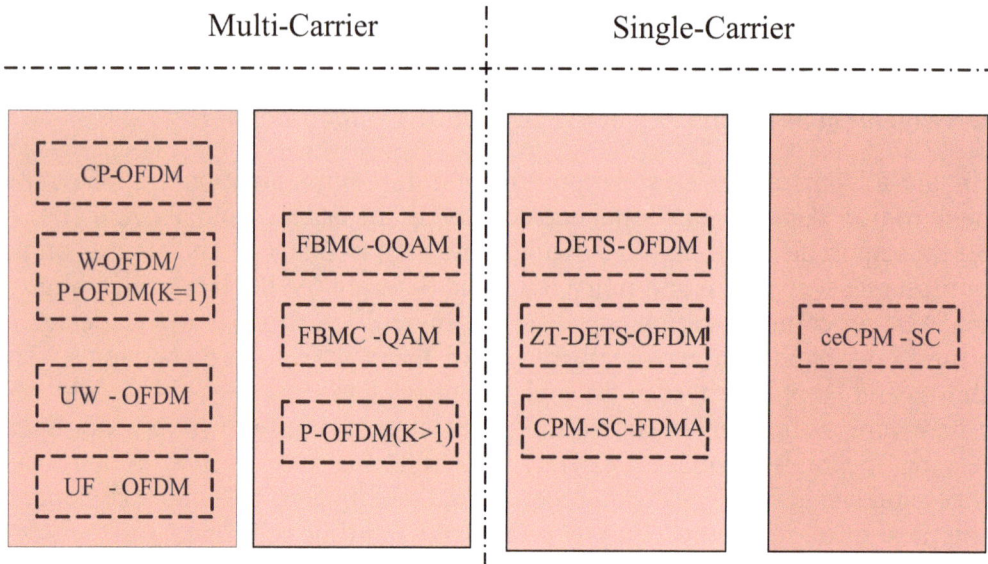

```
┌─────────────────┐ ┌───────────────┐  ┌───────────────┐  ┌───────────────┐
│ ┌ ─ ─ ─ ─ ─ ┐   │ │               │  │               │  │               │
│ │  CP-OFDM  │   │ │               │  │               │  │               │
│ └ ─ ─ ─ ─ ─ ┘   │ │ ┌───────────┐ │  │ ┌───────────┐ │  │               │
│ ┌ ─ ─ ─ ─ ─ ┐   │ │ │FBMC-OQAM  │ │  │ │ DETS-OFDM │ │  │               │
│ │  W-OFDM/  │   │ │ └───────────┘ │  │ └───────────┘ │  │ ┌───────────┐ │
│ │P-OFDM(K=1)│   │ │ ┌───────────┐ │  │ ┌───────────┐ │  │ │ ceCPM-SC  │ │
│ └ ─ ─ ─ ─ ─ ┘   │ │ │FBMC -QAM  │ │  │ │ZT-DETS-OFDM│ │  │ └───────────┘ │
│ ┌ ─ ─ ─ ─ ─ ┐   │ │ └───────────┘ │  │ └───────────┘ │  │               │
│ │ UW - OFDM │   │ │ ┌───────────┐ │  │ ┌───────────┐ │  │               │
│ └ ─ ─ ─ ─ ─ ┘   │ │ │P-OFDM(K>1)│ │  │ │CPM-SC-FDMA│ │  │               │
│ ┌ ─ ─ ─ ─ ─ ┐   │ │ └───────────┘ │  │ └───────────┘ │  │               │
│ │ UF - OFDM │   │ │               │  │               │  │               │
│ └ ─ ─ ─ ─ ─ ┘   │ │               │  │               │  │               │
└─────────────────┘ └───────────────┘  └───────────────┘  └───────────────┘
```

Fig. (3). Classification diagram of 5G candidate waveform.

MC waveform modulation technology includes: CP-OFDM, W-OFDM, P-OFDM, UW-OFDM, UF-OFDM, FMMC/QAM, FBMC/OQAM. SC waveform modulation technology includes: DFT-S-OFDM, ZT-DFT-S-OFDM, CPM-S--FDMA, ceCPM-SC.

In the MC, waveforms can be further divided into two groups, based on: (1) mainly time localized waveforms, with no or little overlap between adjacent waveform symbols; (2) mainly frequency localized waveforms, in which there is a significant overlap between consecutive symbols. The classification is shown in Fig. (3), and the design parameters of all wavcforms are given in Table 3. The waveforms that overlap between adjacent symbols are parameterized by the overlap factor K, which means that the given symbol overlaps the preceding Kmur1 symbol and the subsequent Kmur1 symbol. For FBMC-OQAM and FBMC-QAM waveforms, we mainly study $2 \leq K \leq 4$.

SC waveform modulation technology includes: DFT-S-OFDM, ZT-DFT-S-OFDM, CPM-SC-FDMA, ceCPM-SC. Three of them are based on OFDM, and one waveform is pure single carrier, as shown in Fig. (4). The design parameters of all waveforms are shown in Table 3. The system block diagram of the transmitter and receiver of the above waveforms can be shown in Fig. (4):

```
┌──────────┐   ┌──────────┐   ┌─────┐   ┌──────────┐   ┌──────────┐
│ OAM/PAM  │──▶│Grouping of│──▶│ S/P │──▶│ Zero-tail │──▶│   DET    │
│  /CPM    │   │ Symbols  │   │     │   │ Insertion │   │ Precoding│
└──────────┘   └──────────┘   └─────┘   └──────────┘   └──────────┘
                                                              │
                                                              ▼
                                                        ┌──────────┐
                                                        │ IFFT per │
                                                        │  group   │
                                                        └──────────┘
                                                              ▲
┌──────────┐   ┌──────────┐   ┌──────────┐   ┌──────────┐   │
│  Shift,  │◀──│Group-wise│◀──│Windowing │◀──│ CP/UW/GT │◀──┤
│ Overlap  │   │ Filtering│   │ /Pulse   │   │Insertion │   Time
│   Add    │   │          │   │ Shaping  │   │          │   Domain
└──────────┘   └──────────┘   └──────────┘   └──────────┘   Repetition
```

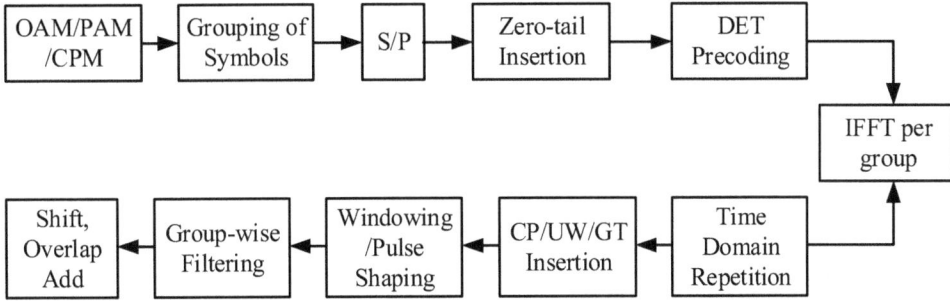

Fig. (4). System block diagram of each candidate waveform transmitter and receiver.

Table 3. Parameter design and configuration table of each waveform.

Waveform	MC/SC	Design Parameter	Synthetic Operation
CP-OFDM	MC	Sub-carrier spacing, FFT Size, CP length	A, C, F, H
W-OFDM	MC	Sub-carrier spacing, FFT Size, CP length, window	A, C, F, H, I
P-OFDM	MC	Sub-carrier spacing, FFT Size, CP length, pulse shape	A, C, F, H, I
UW-OFDM	MC	Sub-carrier spacing, FFT Size, UW (Unique Word)	A, C, F, H
UF-OFDM	MC	Sub-carrier spacing, FFT Size, Subband Filters, Guard time, CP on/off	A, B, C, (D), F, H, J
FBMC-QAM	MC	Sub-carrier spacing, FFT Size, prototype filters and their overlapping factor	A, B, C, F, I, K
FBMC-OQAM	MC	Sub-carrier spacing, FFT Size, prototype filter and its overlapping factor	A, B, C, F, I, K
DFT-s-OFDM	SC	Sub-carrier spacing, FFT Size, number of DFT blocks, CP length	A, C, E, F, H
ZT-DFT-s-OFDM	SC	Sub-carrier spacing, FFT Size, number of DFT blocks, CP length, number of zero-head/zero-tail samples	A, C, D, E, F, H

Fig. **4** quantifies the process of symbol mapping, serial-parallel conversion, Fourier transform, adding CP and filtering in the process of receiving and receiving digital signals. Corresponding to different candidate waveforms, different modules will be selected to build the link between the transmitter and the receiver. The specific parameter configuration and module selection are shown in Table **3**.

Different waveform technologies use component modules with different functions in Fig. (**4**) to build signal processing links, and the system simulation performance varies greatly. As wireless access networks above 6 GHz require huge channel bandwidth, high data transmission rate, and are faced with problems such as complex channel propagation environment and hardware design, relevant parameters are generally used to measure waveforms. A series of indicators such as spectrum efficiency, peak-to-average ratio (PARR), countermeasures against time diversity and frequency diversity, out-of-band leakage, and dynamic parameter configuration are summarized in Fig. (5) [8].

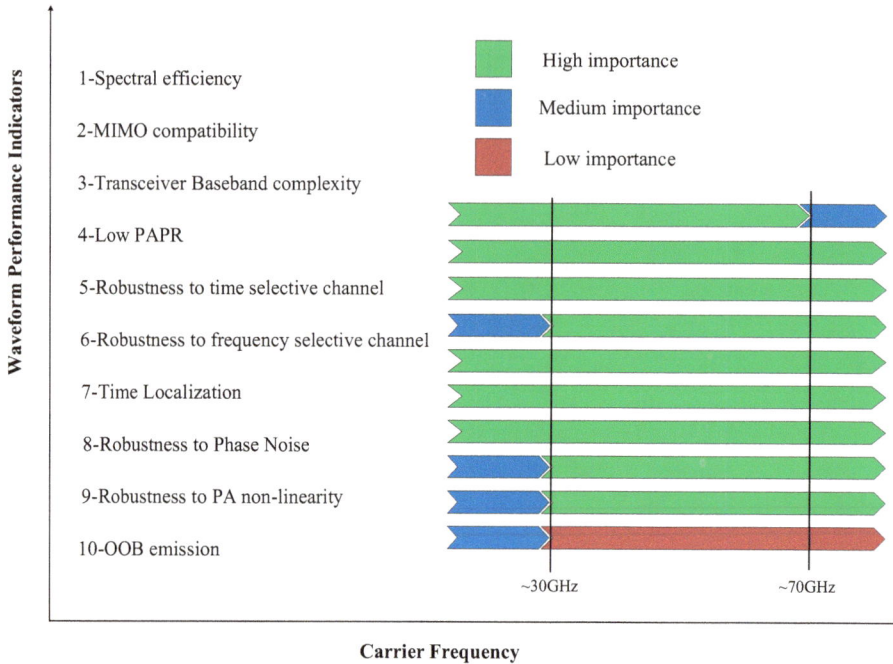

Fig. (5). Schematic Diagram of the importance of Waveform Measurement Indexes in different Carrier Frequency bands.

As can be seen from the above picture, priority should be given to spectrum efficiency, communication capacity, complexity of receiving equipment, time diversity and frequency diversity performance in the submillimeter wave band. In the mmWave band, all the indicators shown in the figure will also cause serious interference to the system performance.

The Influence of RF Link on Millimeter Wave Waveform Signal

The mmWave is the main frequency band of the new generation mobile communication. Considering the characteristics of the new generation mobile communication, the requirements for the performance of mmWave devices in the communication link are also very high. As the carrier frequency increases, the phase noise of the hardware in the link also increases, and the clock offset becomes more significant at higher frequencies, which is a disadvantage for any multicarrier transmission scheme such as OFDM, because these offsets affect orthogonality and make maintaining synchronization more challenging. At the same time, the nonlinearity of RF devices will also have a significant impact on the waveform quality.

Even if the phase-locked loop is used in the 5G communication link working in the mmWave band, it is more likely to be affected by the nonlinearity of radio frequency and the limitation of device fabrication process compared with the low frequency band device. When the modulation mode is high, the phase noise will have a serious impact on the mmWave system. In the 5G mmWave system, the phase noise level of the "total" oscillator (including the reference clock, loop filter and other phase-locked loop components) will be significantly higher than that of the UHF and microwave bands of traditional wireless systems. The phase noise in the oscillator is caused by the noise in the active element and the lossy element which is upconverted to the carrier frequency. A frequency synthesizer usually consists of a reference oscillator (or clock), a VCO (voltage-controlled oscillator) and a PLL with a frequency divider, a phase frequency detector charge pump and a loop filter.

The mathematical characteristics of phase noise are usually described by PSD (power spectral density). The PSD level of the oscillator is about 20 dBc/Hz at ten times the frequency offset of each carrier, which can be expressed as the following formula:

$$S(f) = PSD_0 \prod_{n=1}^{N} \frac{1 + \left(\dfrac{f}{f_{z,n}} \right)^2}{1 + \left(\dfrac{f}{f_{p,n}} \right)^2} \tag{6}$$

where PSD_o is zero frequency ($f = 0$) power spectral density in dBc/Hz, $f_{z,n}$ is zero frequency, $f_{p,n}$ is pole frequency. It can be seen from Equation (6) that as the center frequency of the system rises, the power level of phase noise also rises significantly. At 30 GHz, it can increase by nearly 20 dB compared to 4 GHz. Therefore, the phase noise, which originally had a small effect in the LTE system, has become a major factor that cannot be ignored in the mobile communication system using mmWaves band.

Due to phase noise, the received signal samples will have random time-varying phase errors. In an OFDM system, phase noise can cause CPE (common phase error) and ICI (inter-carrier interference), resulting in the degradation of EVM performance. The phase noise also causes leakage between adjacent channels, which is harmful to the close-range scene of the UE (user equipment) constellation in the cell. Phase noise has an adverse effect on the interference cancellation scheme.

Generally, the phase noise variance increases with the square of the carrier frequency. It is inversely proportional to the power consumption of full frequency generation (PLL). This makes it an important role in mmWave systems, especially those striving to achieve low power consumption, because it may limit throughput. Phase noise also brings phase offset between different symbols. If the center frequency increases, its influence will gradually increase, which will cause great interference to signal decoding. To solve this problem, appropriate methods are needed to track and compensate phase noise.

Phase noise can be compensated in various ways. For example, when the common phase of the constellation rotates in the frequency domain, it is easy to estimate and correct CPE. ICI can be modeled as additional noise. Although it is not always Gaussian noise, it is often difficult to compensate. It requires denser pilots for phase noise and channel tracking, and can be computationally intensive. Generally, the larger the phase noise bandwidth (compared to the subcarrier spacing), the more severe the ICI. Due to the large sub-carrier spacing, only compensation for CPE is sufficient. It is only sufficient to compensate CPE. This should be solved if the estimated channel transfer function of the scattered pilot is used for equalization.

Due to physical limitations, it is generally believed that the efficiency of a PA (power amplifier) in the mmWave band is much lower than that of the PA in the centimeter wave band. In order to achieve a high data rate in the mmWave frequency band, the system bandwidth and modulation order must be increased, and the PAPR of the communication signal will also increase accordingly. Therefore, during the transmission process, the mmWave PA is likely to work in the nonlinear region. If the related technology is used to reduce the PAPR of the system, it will adversely affect the power efficiency of the PA. However, since the PAPR of an OFDM signal tends to increase logarithmically, the PAPR problem is not more serious than the case below 6GHz.

When a signal with a large dynamic range passes through a nonlinear PA, it will be affected by the nonlinearity, resulting in in-band distortion and spectrum regeneration. In-band distortion increases the EVM of the transmitter signal, and spectrum regeneration causes adjacent channel interference. The power series model or polynomial model is widely used in nonlinear PA modeling. The model gives:

$$y(t) = \sum_{k=0}^{K} C_{2k+1} \left| x(t) \right|^{2k} x(t) \tag{7}$$

where K is the nonlinear order, $y(t)$ is the output signal, and $x(t)$ is the input signal, which is the *(2k+1)* complex polynomial coefficient (which can be calculated by the least square method).

Non-linearity of RF link is an important factor affecting waveform quality. To understand this point more deeply, we can refer to the nonlinear amplifier model in Fig. (**6**). The amplifier model shows many types of nonlinear characteristics between transmitter and receiver links. When the output power reaches the compression point, the EVM will increase rapidly.

$$v_0 \left(v_i \right) = v_{DCO} + a v_i - b v_i^2 - c v_i^3 \text{-higher order term} \tag{8}$$

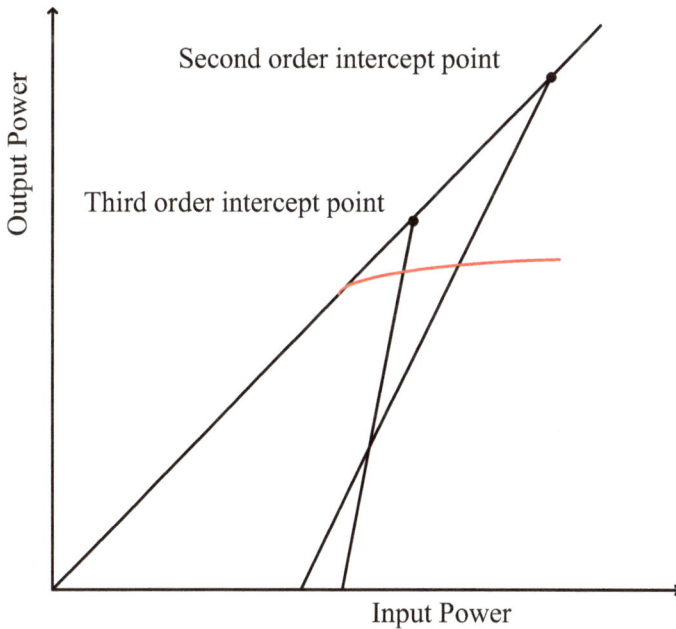

Fig. (6). The working range of nonlinear PA.

$$al\left(1+\frac{\varepsilon}{2}\right)\cos\left(\omega t+\frac{\Delta\varphi}{2}+\theta\right)$$

$$I(t)$$

$$V_{rf}(t)$$

$$Q(t)$$

$$-aQ\left(1-\frac{\varepsilon}{2}\right)\sin\left(\omega t-\frac{\Delta\varphi}{2}+\theta\right)$$

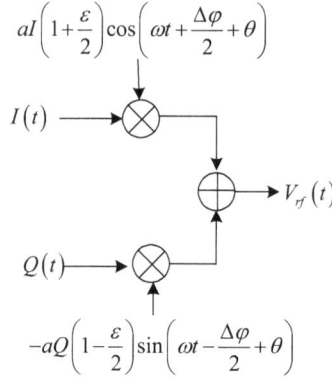

Fig. (7). Schematic diagram of IQ modulator.

When multi-carrier signals with high PAPR pass through nonlinear devices, the signals may be distorted greatly. The industry usually uses 1 dB compression point to specify the nonlinearity of transmitter. The standard radio frequency index of receiver nonlinearity is different from that of transmitter, because the problems to be solved are different. The nonlinear index of receiver is called interception point. The nonlinear relationship between input and output voltage can be estimated by Equation **(8)**, and the DC bias and high-order terms can be ignored in the top-level simulation. As shown in Fig. **(7)**, IQ modulator uses cosine and sine representation of RF carrier frequency. The linear distortion in IQ modulator is caused by the asymmetry of mixer. It always exists in the simulated environment. Asymmetry of mixer can produce gain imbalance A and quadrature error. Quadrature error refers to the phase difference between two local oscillator signals, which has nothing to do with the inherent 90° phase difference. It can produce phase noise and/or frequency offset.

$$v_{rf}(t)=al\left(1+\frac{\varepsilon}{2}\right)\cos\left(\omega t+\frac{\Delta\varphi}{2}+\theta\right)-aQ\left(1-\frac{\varepsilon}{2}\right)\sin\left(\omega t-\frac{\Delta\varphi}{2}+\theta\right) \qquad (9)$$

Massive MIMO for Millimeter Wave Communication

Basic Architecture of Millimeter Wave Massive MIMO

At present, countries around the world are carrying out extensive research on the key technologies of 5G, and the new technologies of 5G may involve many aspects such as physical layer transmission technology and carrier frequency band. At the physical layer transmission level, wireless transmission technology based on massive multiple-input multiple-output (Massive MIMO) can make deep use of spatial wireless resources, and then significantly improve the spectrum efficiency and power efficiency of the system, which has become one of the research hotspots in academia and industry. On the other hand, at the level of carrier frequency band, due to the shortage of spectrum resources in the current cellular frequency band below 6GHz, the implementation of wireless communication in mmWave frequency band has attracted extensive research interest [9]. Due to the relatively high radio wave propagation loss in mmWave frequency band, the research of mmWave wireless transmission technology mostly focused on short-distance communication scenarios in early years, and the related technologies could not be directly applied to large-scale mobile communication scenarios. Considering the relatively short wavelength in mmWave band, large-scale antenna array can be assembled to base station and user side at the same time. Furthermore, the higher beamforming gain provided by the large-scale antenna array can compensate for the relatively higher propagation loss in mmWave band.

With the shortage of spectrum resources, the development and utilization of mmWave spectrum resources on satellite and radar military systems has become the research focus of the new generation of mobile communication technology. Millimeter band is the first choice of Massive MIMO communication system because it has huge space for spectrum resource development. However, due to the short wavelength of mmWave, in mmWave communication, the transmission signal is easily interfered by external noise and other factors, resulting in different degrees of attenuation [10].

Wide deployment of different types of cells (such as macro cells, micro cells, pico cells, and femto cells) to achieve network densification is a key technology to improve network capacity, coverage performance, and energy efficiency. The cell density method has been adopted in existing wireless cellular networks, such as the LTE-Advanced system. Wireless HetNets (Heterogeneous Networks) include remote radio heads and wireless relays, which can further improve network performance. Compared with the existing 4G system, it is expected that relay and multi-hop communication will be one of the basic elements of the 5G wireless architecture. Generally, HetNets' radio resource management plays a vital role in advanced network architecture. Specifically, it is indispensable to develop resource allocation algorithms that effectively use radio resources (including bandwidth, power, and antennas) to reduce inter-cell and inter-user interference and ensure QoS that can be received by active users. In addition, it is very important to design and deploy a reliable backhaul network to achieve effective resource management and coordination, which can be seamlessly integrated with current networks and access technologies. Massive MIMO and mmWave technology provide a crucial means to solve many technical challenges of 5G HetNets in the future.

Deploying a large number of antennas in the transmitter or receiver can significantly improve the spectrum and energy efficiency of wireless networks. At present, most wireless systems operate at microwave frequencies below 6 GHz. In a rich scattering environment, these performance gains can be achieved by simple beamforming strategies, such as maximum ratio transmission or zero forcing. The absolute capacity of the next generation wireless network requires the use of frequency bands above 6 GHz, among which mmWave band can provide huge spectrum resources. Most importantly, the mmWave frequency has a very short wavelength, which can accommodate a large number of antenna elements in a small area, which helps to realize multi antenna connection between base station (BS) and user equipment (UE). MmWave band can be used for outdoor point-t--point backhaul links, supporting indoor high-speed wireless applications (such as high-resolution multimedia streaming). In fact, mmWave communication technology has been standardized for short distance service in IEEE 802.11ad.

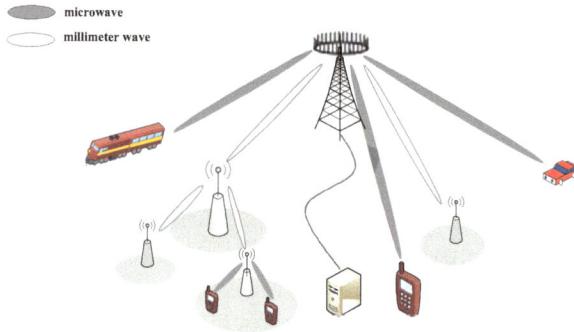

Fig. (8). Schematic diagram of Massive MIMO network based on millimeter.

In the future, the mobile communication system will deploy small cell (such as Pico and FeMo) on a large scale. Short distance mmWave communication technology is very fundamental. The mmWave communication technology is considered as one of the potential technologies to meet the needs of the new generation mobile communication network [11]. There are many different architectures for mobile communication wireless HetNets based on mmWave Massive MIMO. The architecture combining mmWave and microwave is adopted, as shown in Fig. (**8**).

Current Research Direction

There is relatively little work on theoretical and actual modeling of broadband massive MIMO channels in the mmWave frequency band, and the mmWave massive MIMO channel model has rarely been reported.

The mmWave channel in the mobile scene has a serious Doppler effect, and there is a bottleneck problem in the acquisition of instantaneous channel information; most of the transmission methods reported in the literature are based on the ideal assumption of accurately acquiring instantaneous channel information, and there are problems such as high implementation complexity.

Statistical channel information changes more slowly than instantaneous channel information. Using statistical channel information can assist mmWave massive MIMO wireless transmission design and improve system transmission performance.

The mmWave massive MIMO wireless channels exhibit energy concentration and other characteristics in the beam domain. Implementation of millimeter-wave MIMO wireless transmission in the beam domain can alleviate the high path loss of the millimeter-wave channel, and make deep use of spatial wireless resources to achieve multi-user sharing of spatial wireless resources.

It can be seen that the research on mmWave massive MIMO wireless transmission technology is still in its infancy, and there are challenging basic theories and key technical issues [12]. In order to fully explore its potential technical advantages, it is necessary to explore the millimeter-wave massive MIMO channel model in typical mobile scenarios, and analyze its system performance under constraints such as actual channel model, moderate pilot overhead, and implementation complexity. The optimal transmission technology should be explored to solve the bottleneck problem of channel information acquisition involved in millimeter-wave massive MIMO wireless transmission, system implementation complexity, and applicability issues in typical mobile scenarios. Millimeter-wave massive MIMO wireless transmission can expand the utilization of new spectrum resources, and can deeply explore spatial dimensions of wireless resources, and greatly increase the wireless transmission rate. It is one of the most promising research directions to support future broadband mobile communications.

Hybrid Beamforming in MIMO

Beamforming technology is one of the key technologies in wireless communication. Generally speaking, beamforming technology is a communication technology that transforms a transmitting beam into a narrower transmitting beam in real time at the transmitting end, so as to aim the energy at a specific target user and obtain additional system gain to improve the signal quality of a specific target user. Beamforming technology has the following significant advantages:

Longer coverage distance: The power of the wireless unit is unevenly distributed in space, and the main radiation energy only points to the location of the terminal, so as to avoid power waste, improve the overall system gain, and expand coverage distance.

Stronger anti-interference capability: Based on unbalanced power distribution, the antenna array is accurately pointed to the location of the terminal at the transmitting end to minimize the interference to other wireless systems. Moreover, the possibility of interference from other wireless communication systems is also minimized at the receiving end.

Higher transmission rate: Due to the improved quality of the received signal of the system and the higher SNR due to the minimum interference, the system can stably work in higher order modulation mode. Therefore, under the condition of unit spectrum resources, the spectrum utilization ratio is higher and the transmission rate can also be significantly improved.

According to the characteristics of weight vector, the beamforming technology can be divided into two schemes: fixed weight beamforming and adaptive beamforming. Weights fixed beam informs refers to that the system of the beam weight vector and the direction of the antenna diagram are fixed. The fixed beam can form different beam combination weighting vector. The base station side choose different combination to launch out and choose the make in receiving the combination of the corresponding maximum signal-to-noise ratio at the receiving end based on the beam search methods. According to the different paths of the user signal propagation in space, the adaptive beamforming is optimized to form the direction graph, and different antenna gains are given in different arrival directions to form the narrow beam aiming at the user signal in real time. The side lobe is as low as possible and the directional reception is adopted to improve the capacity of the system in other directions. Due to the user's mobility and scattering environment, the arrival direction of the signals received by the base station is time-varying. Adaptive beamforming can be used to separate the signals with similar frequency but separable space [13].

The mmWave frequency band has a large amount of available spectrum resources, but at the same time, the transmission characteristics of millimeter-wave, such as severe path loss and poor channel quality, make the implementation of millimeter-wave technology face some difficulties. In addition, in the mmWave MIMO system, if digital beamforming technology is used, the number of RF links will increase. Meanwhile, the increasement of the cost and power consumption is considerable for the millimeter-wave communication system.

The hybrid beamforming technology could be utilized to solve the problem of too many RF links and serious path loss. In general, the prohibitively high cost and power consumption discourage the use of digital beamforming for mmWave massive MIMO systems. Hybrid beamforming, with less RF chain than the number of antennas, is feasible for mmWave massive MIMO and exhibits only a negligible performance loss when compared to the digital beamformer, which though has optimal performance for conventional MIMO systems but practically infeasible for mmWave massive MIMO systems.

Hybrid beamforming greatly reduces the use of RF chain and has greater advantages in mmWave MIMO systems. In the hybrid beamforming system shown in Fig. (9), the baseband data flow is first processed through the digital domain to control the amplitude, and then sent through a limited number of RF chains to the analog domain for processing to control the phase.

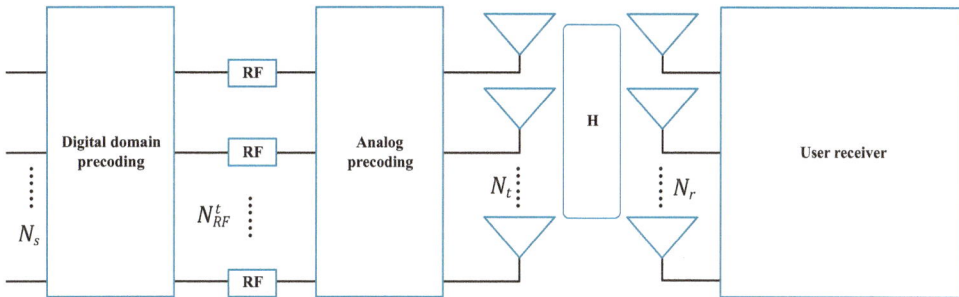

Fig. (9). Block diagram of hybrid beamforming.

Hybrid precoding structure is a way to provide stronger gain for MIMO communication in millimeter-wave frequency band. The existing hybrid precoding system architectures are mainly divided into two categories, namely the mainstream hybrid precoding structure based on phase shifter introduced above and the hybrid precoding structure using low power consumption switch structure which is less applied. Among them, mixed precoding structure using phase shifter network can use finite resolution phase shifter instead of high-resolution phase shifter. Although the energy consumption of phase shifter network can be reduced without causing obvious performance loss, the structure still has a large amount of power consumption due to the presence of a large number of phase shifters. The switching structure-based precoding does not use phase shifter network, but switch network, which reduces hardware cost and power consumption significantly, but has obvious performance loss. Therefore, the main research direction of hybrid precoding is to study the hybrid precoding structure with high energy efficiency, which can significantly reduce the complexity of the system and maintain the system's good accessibility rate performance.

Spatial multiplexing can be used to drastically boost the system capacity of mmWave massive MIMO channels as the number of BS and UE antennas is much larger than in conventional systems. It increases transmission throughput by subdividing outgoing signals into multiple streams where each stream is transmitted simultaneously and in parallel on the same channel through different antennas. At mmWave frequency bands, signals are easily diffracted by physical objects thereby leading to several reflected signals from the different scatters. Such rich-scattering environment favors independent and parallel data streams which increase spatial multiplexing gains [14].

However, mmWave channels are specular and have low rank, particularly for line of sight (LOS). They tend to have a lower number of time clusters and spatial lobes (or generally less multipath components) compared to μWave. They are thus incapable of exploiting all available DoF (*i.e.*, only a couple of spatial streams can be supported), thereby limiting achievable multiplexing gains. Also, sufficient decorrelation between different closely-spaced antennas needed by the channels for optimal spatial multiplexing is usually unrealizable in most practical systems such as mmWave indoor.

Millimeter Wave Network Backhaul Technology in Mobile Communication

Millimeter Wave Backhaul Network Architecture

The mmWave backhaul network is composed of macro base stations, several micro base stations and randomly distributed users. The macro base station accesses the core network through the gateway. Micro base stations are connected to each other, and connected to the macro base station to access the core network. Users can communicate with both macro base stations and micro base stations. The transmission link between the base station and the base station is called the backhaul link, and the transmission link between the base station and the user is called the access link [15]. A typical mmWave backhaul network architecture is shown in Fig. (**10**).

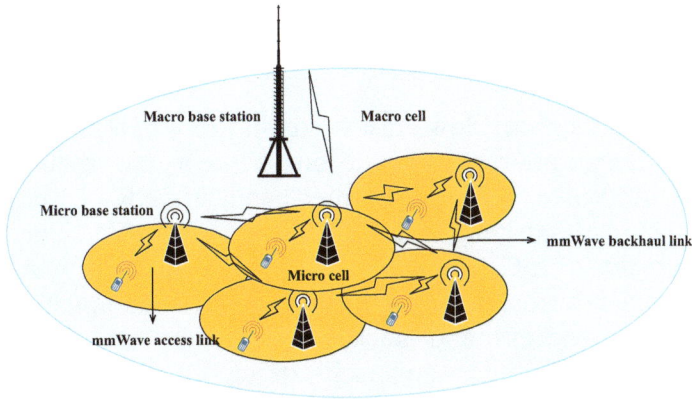

Fig. (10). MmWave backhaul network architecture diagram.

Millimeter Wave Network Backhaul Technology

As one of the key technologies in the new generation of mobile communications, backhaul technology plays an unparalleled role. Here we will focus on the 5G NR (New Radio) technology, self-backhaul technology and small base station backhaul technology. Self-backhaul technology has played a vital role in the 5G NR integration scheme, and small base station backhaul technology is crucial to the future deployment of small base station.

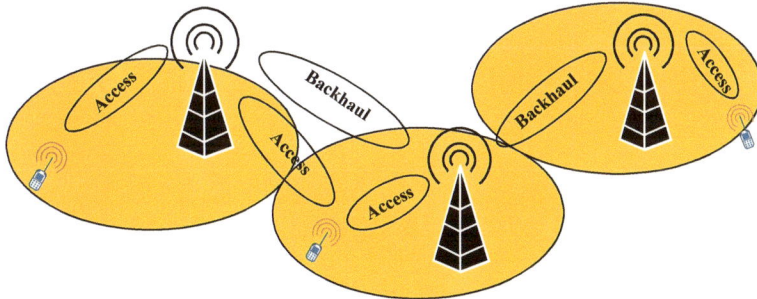

Fig. (11). Schematic diagram of integrated wireless access and backhaul scheme.

5G NR Technology

In March 2017, 3GPP held a meeting to pass the proposal for 5G NR standard acceleration and proposed how to solve the 5G NR integrated wireless access and backhaul problems. To achieve ultra-dense networking, further integration of access and backhaul links is the key to the problem. In addition, 5G NR requires a wider bandwidth than 4G LTE network. Especially in the mmWave frequency band, the deployment of additional Massive MIMO and multi-beam systems has promoted the solution of 5G NR integrated wireless access and backhaul. As shown in Fig. (**11**), in a network that integrates wireless access and backhaul links, multiple wireless access and backhaul links are supported, making the deployment of 5G NR cells more integrated.

An integrated solution that allows fast switching of wireless access and backhaul links is particularly important. On the one hand, it can reduce the short-term congestion of mmWave, and on the other hand, 5G NR can be deployed more flexibly. It is shown that integrated wireless access and backhaul determine the initial deployment of 5G networks. Once 5G NR equipment is launched, integrated wireless access and backhaul technologies must keep up. If the integrated wireless access and backhaul solutions are delayed, the timely deployment of 5G NR will be adversely affected.

Self-Backhaul Technology

Self-Backhaul technology originates from the existing 4G LTE protocol, which uses LTE wireless resources to implement layered-three point-to-point relay between eNB (Evolved Node B), requiring a collaborative schedule between relay eNB and anchor eNB. The working principle of self-backhaul technology is shown in Fig. (**12**). Due to the densification of mobile communication networks in the future, the traditional CPRI (Common Public Radio Interface) fronthaul cannot support the scale of mobile networks and considering the cost of network construction. The advantages of self-backhauling technology are highlighted, and the self-backhauling technology solutions are more urgency and thoroughness. If this scheme can be promoted on a large scale, the design, maintenance and optimization of the future mobile communication network will undergo tremendous changes. Moreover, due to the demand to optimize routing, balance wireless access and return traffic ratios, as well as to achieve self-configuration and self-healing ability of network backhaul, plenty of software, machine learning, artificial intelligence and other cross-cutting technologies will be introduced.

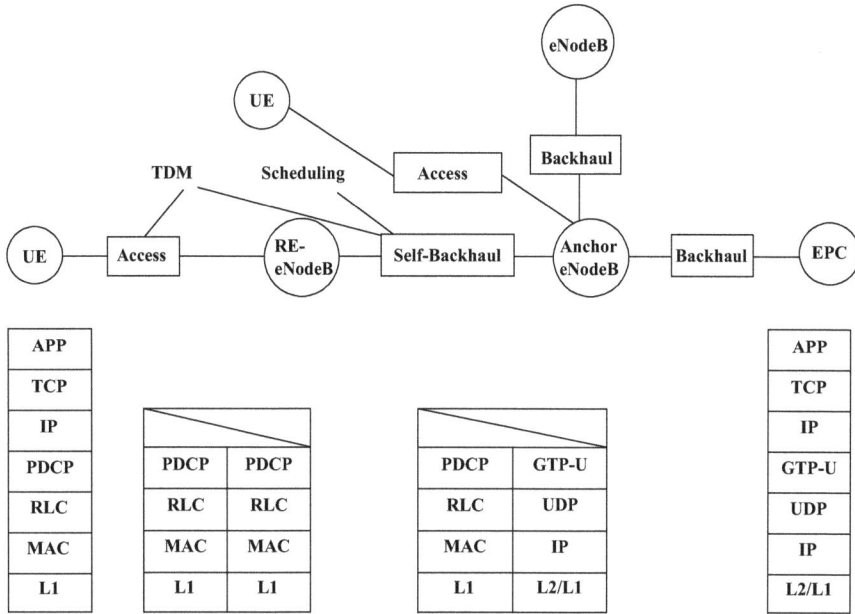

Fig. (12). Schematic diagram of the work principle of self-backhaul.

Small Base Station Backhaul Technology

At present, with the improvement of 4G networks and the large-scale deployment of NB-IoT (Narrow Band Internet of Things) networks, small base stations such as Small Cell and Book RRU are deployed on a large scale in urban areas for network blindness compensation, heat absorption and traffic distribution. It is foreseeable that the demand for high-speed and ultra-dense deployment in the future communication era will be more vigorous. Furthermore, the site deployment density will further increase and small base station deployment will become the mainstream deployment method. From the perspective of network deployment structure, macro cell base stations are mainly used for network layout. From their characteristics, macro cell base stations are relatively sparse and require high access reliability. The main consideration is to use PTN (Packet Transport Network) technology to access. Currently, small base stations mainly consider the use of PON (Passive Optical Network) technology in the access layer, and it is connected to the IP packet network and related core network through the BRAS (Broadband Remote Access Server). In order to meet the frantically growing demand for data services, the ever-increasing miniaturization and intensification of massive small base stations have put forward higher and more demanding requirements for data backhaul performance indicators [16].

The industry is still actively exploring wireless backhaul methods with reliable access, low cost, and flexible deployment. In addition to the aforementioned 5G NR technology, self-backhaul technology, and small base station backhaul technology, Facebook's Terragraph network is also a good attempt. It replaces the existing urban street lamp network with base station equipment of wireless signal transmission function to form the mmWave mesh wireless backhaul network, thus replacing the traditional expensive and time-consuming optical fiber access method. Moreover, AT&T's AirGig project considers installing and deploying mmWave equipment on electric poles, using ubiquitous power lines as the carrier to form a powerful wireless backhaul network and to achieve seamless coverage of indoor and outdoor signals. Moreover, the F-Cell technology of Bell Labs uses Massive MIMO beams to form a self-organizing wireless backhaul architecture whitout the traditional optical fiber backhaul. The large-scale and flexible deployment of network Small Cells are realized to form a unique base station wireless backhaul the way.

Key Technologies for mmWave Network Backhaul

Effective Spatial Reuse

The ultimate goal of backhaul transmission is to improve throughput. Compared with traditional wireless cellular networks, the interference caused by highly directional transmission in mmWave communication is much smaller, paving the way for spatial multiplexing, effectively saving frequency and time resources, and improving throughput. Spatial reuse is particularly suitable for dense cell scenarios, because densely distributed devices bring greater opportunities for parallel transmission. TDMA (Time Division Multiple Access) is used in the IEEE 802.11ad standard, and STDMA (Spatial TDMA) has recently attracted widespread attention. The transmission is performed in units of time intervals of equal length, called time slots. The purpose of scheduling is to allocate as many links as possible in the same time slot to make full use of space multiplexing. At present, most scheduling studies related to mmWave are based on half-duplex links in a single-user scenario, that is, adjacent links with common nodes cannot transmit at the same time, and devices can only send or receive a single data stream at a time. However, the development of physical layer technology in recent years, such as full duplex, hybrid beamforming, and multicast beamforming may change scheduling to further increase the spatial multiplexing gain [17].

Efficient Path Establishment

The small base station based on the wireless self-backhaul technology needs to have a plug-and-play function. After the small base station is deployed and powered on, it needs to find a host base station to connect to the wireless access network. In order to save the energy consumption of the base station, the small base station which has just turned on actively initiates a discovery request, and the base station with the host function responds after receiving the request. The small base station selects an optimal base station as its donor base station based on comprehensive factors such as the received signal strength, distance, and available transmission resources and service load contained in the response message. In order to ensure the robustness of the wireless backhaul, one optimal can be selected as the primary donor base station, and several other sub-optimal ones can be selected as secondary donor base stations. The path establishment of the wireless self-backhaul network is to find an optimal path from a small base station to a macro station.

The establishment of the wireless self-backhaul network path can adopt both centralized and distributed methods. In the centralized method, the access network needs to have a centralized control node, which can be a macro base station, or its functional equipment can be developed separately. The centralized control node obtains the interconnection information between the base stations, performs overall planning of the wireless self-backhaul network path, and notifies the results to each base station. This method can obtain the overall optimum, but it is not flexible enough and the response is slower. Distributed path establishment does not require the development of additional centralized control functions. Each small base station obtains the path information of the wireless neighbor base station and makes a local optimal selection. Although this method may not be able to obtain the overall optimum, it is fast, flexible, and low in development cost.

In order to improve the transmission bandwidth and transmission performance of the wireless backhaul link, a multi-connection path can be established in the wireless self-backhaul network, and the small base station is connected to the macro base station through the multi-hop path. In the multi-connection transmission mode, the problem that needs to be solved is the distribution functions of control plane and user plane, and the balanced distribution of user plane bearer resources on different connection paths.

Business requirements and wireless transmission link conditions may change during operation between small base stations. In addition, small base stations have the need for nomadic deployment. For consideration of environmental protection and energy conservation, small base stations may also be constantly switched on and off. Therefore, the wireless backhaul link needs to be updated in real time. The premise of path update is the real-time accurate measurement and transmission of the wireless backhaul link transmission conditions.

Reasonable Resource Allocation

Wireless link transmission is the main factor affecting the performance of mobile communication networks. For an ultra-dense wireless access network based on wireless self-backhaul technology, the transmission quality of the access link and the backhaul link will directly affect the user experience. Therefore, when performing wireless resource allocation, it is necessary to comprehensively consider user business requirements and transmission road quality, joint design of access link and backhaul link and joint resource allocation.

The backhaul link and the access link can be deployed at the same frequency (in-band mode) or at the different frequency (out-of-band mode). Co-frequency deployment is relatively simple, but the backhaul link occupies part of the system transmission resources and reduces the user transmission rate. Inter-frequency deployment is more complicated, and the radio frequency of a small base station must have multiple sets of radio frequency transceiver devices. However, it can obtain greater spectrum efficiency and user transmission rate. In addition to spectrum bandwidth resources, base station power, CPU processing capacity and number of antennas are also system resources that need to be jointly allocated. System resource allocation is to allocate appropriate access resources to different users and allocate appropriate backhaul resources to different small base stations. If there are too many access resources and too few backhaul resources, the backhaul link will become a transmission bottleneck for wireless access. If more resources are allocated to the backhaul link, the transmission quality and system capacity of the access link cannot be guaranteed, and the utilization rate of the backhaul link will not be very high. Therefore, system wireless resource allocation is to find the best balance point between access resources and backhaul resources, so that system capacity, spectrum efficiency, and power efficiency can be optimized at the same time.

Convenient Access Network Cache

In order to meet the end-to-end low latency requirements of mobile networks, MEC (Mobile Edge Computing) technology is introduced. MEC is a network architecture that provides services required by users and cloud computing functions on the wireless side. MEC is used to accelerate the rapid download of various applications in the network and allow users to enjoy an uninterrupted high-quality network experience. MEC requires the deployment of small base stations with local computing and storage capabilities at the edge of the mobile network. One of the safeguards is that service content can be cached in the small base stations of the wireless self-backhaul network. Caching was originally proposed in content distribution networks to enhance data locality, that is, through content replication at strategic nodes in the network (such as proxy servers, gateways), and at the same time balance network traffic during off-peak hours. After the access network cache is implemented, the hot file or the same file requested by the user can be directly obtained from the cached base station without accessing the core network. This greatly reduces the network delay, and the load of the backhaul link can also be reduced, thereby reducing the demand for backhaul resources, improving the allocation of access resources, and increasing the capacity of the access network system. Essentially, by separating the time for downloading the file content from the time for delivering it to the user, the small base station can improve the quality of service experienced by the user while saving backhaul resources.

Millimeter Wave Network Technology in Mobile Communications

Ultra-Dense Network

In order to meet the needs of future mobile network data traffic increase by 1000 times and user experience rate increase by 10 to 100 times, in addition to increasing spectrum bandwidth and using advanced wireless transmission technology to improve spectrum utilization, the most effective way to increase wireless system capacity is still increase spatial reuse through encrypted cell deployment [18]. Traditional wireless communication systems usually use cell splitting to reduce the cell radius. However, as the cell coverage is further reduced, cell splitting will be difficult. It is necessary to intensively deploy low-power small base stations in indoor and outdoor hotspots to form UDN.

UDN is an effective solution to solve the explosive growth of future mobile communication network data traffic. It is predicted that in the area covered by macro base stations in the future wireless network, the deployment density of low-power base stations with various RAT (Radio Access Technology) will reach more than ten times the density of existing sites, to form ultra-dense heterogeneous network.

Key Technology of Network

Virtual Layer Technology

UDN uses virtual layer technology, that is, a single-layer physical network builds a virtual multi-layer network. As shown in Fig. (**13**), single-layer physical micro base station cell builds two-layer network (virtual layer and physical layer), and the macro base station cell serves as the virtual layer. The virtual macro cell carries control signaling and is responsible for mobility management. The physical micro base station cell serves as the physical layer, and the micro cell carries data transmission. This technology can be realized through a single-carrier or multi-carrier scheme. The single-carrier scheme constructs a virtual multilayer network through different signals or channels. The multi-carrier scheme constructs a virtual multilayer network through different carriers, and combining multiple physical cells (or multiple physical part of the resources on the cell) is virtualized into a logical cell. The resource composition and settings of the virtual cell can be dynamically configured and changed according to the user's movement and business needs. The virtual layer and user-centric virtual cells can solve the mobility problem in ultra-dense networking.

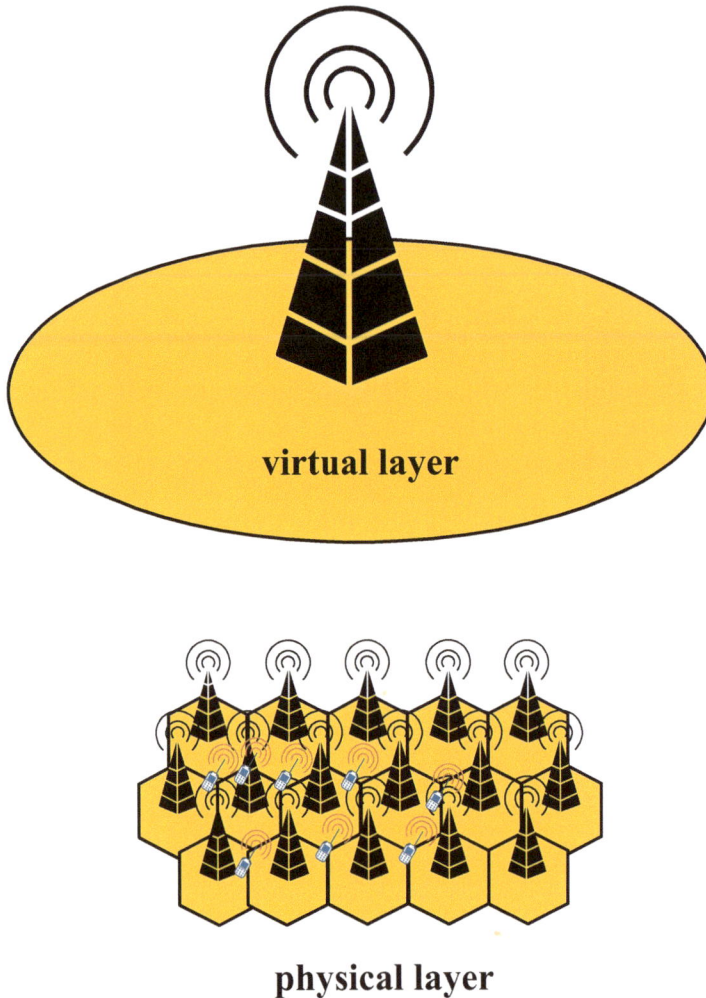

virtual layer

physical layer

Fig. (13). Schematic diagram of virtual layer technology principle.

Multi-Connection Technology

For macro-micro heterogeneous networking, micro base stations are usually deployed locally in hotspots, and there may be holes with discontinuous coverage in the middle. Based on this architecture, it is not only necessary to have the control plane function corresponding to the signaling base station from the perspective of the macro base station, but also necessary to carry user plane data in the deployment area of the micro base station according to actual deployment requirements. The fundamental purpose of adopting multi-connection technology is to realize simultaneous connection between user terminals and multiple network nodes. In different network nodes, the same access technology can be used, or different access technology can be used. Since the macro base station does not need to process the user plane corresponding to the micro base station, strict synchronization between the macro base station and the micro base station is not required, which greatly reduces the requirements for link performance. In the dual link mode, the macro base station is the main base station, which can provide a unified and centralized control plane, while the micro base station is the secondary base station, which can only carry user plane data and cannot connect to the user terminal control plane. After the primary and secondary base stations complete the negotiation for radio resource management, the secondary base station transmits the configuration information to the primary base station on the interface, and the radio resource control message is only sent by the primary base station, and finally reaches the user terminal. The radio resource control entity of the user terminal can see a message sent from the radio frequency unit entity, and the user terminal can only respond to the radio resource control entity. The user plane is not only distributed among micro base stations, but also exists among macro base stations. Because the macro base station can provide similar functions as the data base station, it can effectively solve the transmission problem of the micro base station covering the unconnected parts.

The current wireless backhaul works in line-of-sight (LOS) propagation environment, microwave frequency band or mmWave frequency band, and the actual propagation rate is about 10 Gbit/s. There is a big difference between the current wireless backhaul and the existing access technology in technical methods and resources. In the current existing network architecture, the horizontal communication between base stations still has a very high delay, and the base stations cannot be plug and play. Both network deployment and network maintenance require high costs, which are mainly affected and restricted by the actual conditions of the base station itself. The bottom layer backhaul network cannot provide reliable support for this function. In order to make node deployment as flexible as possible and reduce deployment costs, the spectrum and technology that are completely consistent with the access link can be used to achieve wireless backhaul transmission. For wireless backhaul, wireless resources can not only provide services for terminals, but also provide relay services for nodes.

Anti-Interference Management

For high-definition video, AR/VR, drones, industrial robots, AI and other application scenarios, high speed, high density, and large capacity are important capabilities of mobile communication networks, and cell miniaturization and densification are the main features of UDN, which can greatly improve the spectrum efficiency and the capacity of the access network system, thereby providing users with high-speed Internet experience. However, the densification of network nodes makes inter-cell interference serious, result in co-channel interference, shared spectrum interference, interference between different coverage levels. The transmission loss of neighboring nodes has little difference, and there are multiple interference sources information with similar strengths.

Interference coordination is a technology with more applications. Since LTE adopts OFDMA (Orthogonal Frequency Division Multiple Access), it can effectively avoid the interference of users in the cell. Therefore, the interference coordination mainly refers to the ICIC (Inter-Cell Interference Coordination). The basic idea is that the cells in the area coordinate the scheduling and allocation of resources according to the set rules and methods, thereby reducing mutual interference and improving transmission quality. According to the coordination period, interference coordination can be divided into static interference coordination, semi-static interference coordination and dynamic interference coordination. At present, interference coordination between static cells is more commonly used, and its implementation is simple and convenient. The other two methods require information exchange (X2 interface) and strict delay requirements. Under the premise that the equipment algorithm is mature enough and the network planning is feasible, it can not only bring more optimized performance, but also minimize manual intervention.

Wireless Return Method

Due to the ultra-dense and large-capacity characteristics of UDN, the backhaul format needs to be changed. Traditional wired backhaul is no longer applicable due to the difficulty of deployment and high cost. Therefore, the wireless backhaul method using the same spectrum as the access link is more important.

In addition, from the perspective of UDN network, plug-and-play is also the direction of future micro-station evolution. Therefore, if only considering the way of fiber cable backhaul, it will obviously restrict the deployment of a large number of small and micro-stations. In the era of shared resources, the deployment of base stations has become more flexible, and the way of backhaul has also changed accordingly. Table 4 shows the recommended backhaul methods in different scenarios. It can be foreseen that smart buildings such as dense residential buildings, dense neighborhoods, commercial office buildings, large venues and clubs, temporary meeting places, universities, subways, and emergency rescue and disaster relief command sites may all have wireless backhaul requirements.

Table 4. Return methods in different scenarios.

Application Scenario	Classification	Scenario Characteristics	Return Method
Office building	Indoor	Rich site resources, sufficient transmission resources, static or slow movement of users	Wired backhaul
Stadium	Indoor	Rich site resources, sufficient transmission resources	Wired backhaul
Subway	Indoor	Dense and mobile users	Wireless backhaul
Assemblies	Indoor and outdoor	Dense users are stationary or moving slowly	Wireless backhaul
Residential area	Indoor and outdoor	Users are stationary or moving slowly	Wireless backhaul
Campus	Indoor and outdoor	Dense users are static or slow moving	Wired/wireless coexistence

From the perspective of wireless backhaul design, mobile networks also propose many possibilities. For example, the backhaul link and the access link may be deployed at the same frequency or at different frequencies. Ultra-dense micro-stations have higher requirements for site locations, which are mainly reflected in the requirements for transmission resources. The plug-and-play requirements of micro-stations make wireless backhaul an effective way to solve the limited transmission resources, which is easy and flexible to deploy.

Problems Faced by Network

In hotspots with high-capacity intensive scenario, multiple wireless environments are very complex and easily affected by interference. UDN of base stations can effectively improve the spectrum efficiency of the system itself. Moreover, rapid and reasonable deployment of wireless resources can also be realized with the rapid scheduling of resources, and existing wireless resources can be fully utilized on the basis of improving the actual spectrum efficiency of the system. However, this also brings many practical problems, mainly including the following aspects:

System Interference

In a dense, complex and heterogeneous scenario, different access sites coexist, which may cause system interference. If the interference reaches a certain level, the system's spectrum efficiency will seriously deteriorate.

As the distance between wireless access sites continues to decrease, handovers between cells must become more frequent, leading to a significant increase in the actual consumption of signaling. The user's own business service capabilities are ultimately reduced, and service quality cannot be guaranteed.

System Cost and Energy Consumption

In order to effectively meet the throughput requirements in hotspot areas and provide users with a new experience, many wireless access nodes must be introduced. At the same time, most of these wireless access nodes are very dense, making frequency resources richer, and the types of access technology also increased. In these cases, system deployment will consume a lot of cost and energy.

Low-Power Base Station

In order to enable low-power base stations to achieve rapid and flexible deployment, plug-and-play functions are required, including autonomous backhaul functions, automatic configuration functions, *etc*.

UDN Deployment Plan

The ultra-dense network deployment breaks the traditional flat single-layer network structure and makes the emergence of a multi-layer HetNet. The ultra-dense micro base stations have become the key in the HetNet architecture of mobile communications. The proportion of traffic carried by small base stations will increase significantly with the large-scale deployment of ultra-dense micro base stations. The deployment plan and deployment architecture of the network are discussed.

Deployment Plan

The network plan in the new generation of mobile communications is mainly aimed at business network forms such as wide coverage, hot spots high capacity, low latency and high reliability, and large-scale MTC (Machine-type Communications). The characteristics of each form are as followed. The network form of the business scenario is mainly based on the coverage of the macro cell base station cluster. The core network control function is centrally deployed with hosting high mobility, and the radio resource management function is submerged to the macro cell and base station cluster. For the network form of hotspot high-capacity business scenarios, microcells supplement the hotspot capacity while combining wireless technologies such as large-scale antennas and high-frequency communications. The core network control plane is deployed in a centralized manner. In the macro-micro and micro cell cluster scenarios with severely limited interference, collaborative resource management and small-scale mobility management sink to the wireless side, and user plane gateways, service enablement and edge computing sink to the access network side to realize local services streaming and rapid content distribution. For the network form of low-latency and high-reliability business scenarios, control functions and large-scale mobility-related functions are concentrated, and small-scale mobility management functions and specific business-specific control functions are lowered to the wireless side. The gateway, content caching, and edge computing sink to the wireless side to realize fast service termination and distribution, and support direct communication between network-controlled devices. For the network form of large-scale MTC business scenarios, the network control function is customized and tailored according to the MTC business, and MTC information management, policy control, MTC security are added. Moreover, general control modules such as mobility management are simplified. The user plane gateway is sinking, and convergence is increased. The gateway realizes network access and data aggregation services for a large number of terminals, and provides network connection services based on coverage enhancement technology in weak coverage areas and blind areas.

The important development direction of the new generation of mobile communication planning coverage is refined ultra-dense network. According to the requirements of different scenarios, it adopts multi-system, multi-layer, multi-cell, and multi-carrier networking to meet the needs of different business types.

Deployment Architecture

In UDN, the heterogeneous architecture of macro base stations and micro base stations can be used for deployment to improve the network's own traffic density and peak rate. In addition, in order to meet the requirements of different performance indicators, the distance between base stations will be further shortened. The development and utilization of resources in different frequency bands and different wireless access technologies will eventually form a network architecture which fully meets the requirements.

Macro base station & Micro base station. In this planning and deployment architecture, from a service perspective, macro base stations transmit low-speed and high-mobility services, while micro base stations mainly provide high-bandwidth services. The above functions achieve the goal of allowing macro base stations to cover and resource management between micro base stations, that is, micro base stations are mainly responsible for capacity. Thus, the access network can be deployed flexibly based on business development requirements and combined with distribution characteristics, and finally realize the separation between control and bearer. When the separation between control and bearer is achieved, the actual problem of multiple handovers under dense networking conditions is effectively solved, and the user experience is improved while resources are more fully utilized.

Micro base station & Micro base station. This mode does not introduce macro base stations. In this planned deployment mode, in order to achieve functions similar to the above modes, it is necessary to construct a virtual macro group. In this process, different micro base stations in the cluster need to share some resources, such as carriers, signals, and channels. Different micro base stations in the same cluster can realize the virtualization of the macro cell by transmitting the same resources. At the same time, different micro base stations implement separate transmission of their remaining resources, thereby realizing the separation between the data plane and the control plane. When the network load is low, the micro base stations are managed in clusters, and the micro base stations in the same cluster are combined with each other to form a virtual macro base station and send the same data. Based on this actual situation, the terminal can obtain receiving diversity gain to ensure the quality of the received signal. In the case of a high network load, each micro base station can be used as an independent cell to send corresponding data information to achieve the goal of cell division and achieve a substantial increase in network capacity.

MAIN COMPONENTS OF MILLIMETER WAVE COMMUNICATION

Spectrum is the medium and carrier of wireless signal transmission, and its importance to mobile communication is self-evident. Especially with the approaching of 5G commercialization, how to plan and use high-frequency spectrum, especially mmWave frequency band, has become a hot topic in the industry, and it has also become the key for countries to seize the strategic commanding heights of 5G and even 6G.

The mmWave high frequency band has abundant spectrum resources and outstanding performance advantages. However, because high frequency bands were rarely used in the field of civil communications, the supporting links of the related industry chain were not very mature. The most prominent problem was that the mmWave devices were relatively weak.

In order to overcome the impact of high-frequency key components on system performance, on the one hand, it is necessary to promote the common progress of the industry chain. On the other hand, the overall performance can be improved by optimizing the overall system design, and at the same time, the high-frequency signal path loss can be solved through technical means and networking. And penetration loss issues, such as beamforming, distributed MIMO, and multiple connections.

Filter

The traditional microstrip filter shows problems such as large insertion loss and serious electromagnetic leakage at high frequencies. For mmWave transmitters, the transmit output power is restricted by the chip itself, and the efficiency is very low. The filter with large insertion loss further reduces the output power. In addition, the large insertion loss directly deteriorates the noise figure of the receiver and reduces the receive sensitivity.

The Substrate Integrated Waveguide (SIW) filter is a technology that has attracted much attention in recent years. Its principle is similar to the one of a traditional rectangular wave guide. It has the advantages of low radiation, low loss, and high Q value. At the same time, compared with traditional rectangular waveguide filters, SIW filters are small in size, low in cost, easy to fabricate, and easy to interconnect with other planar devices. They form a system on a PCB to achieve high integration and miniaturization [19]. Thus, SIW filter is very suitable for mmWave communication system.

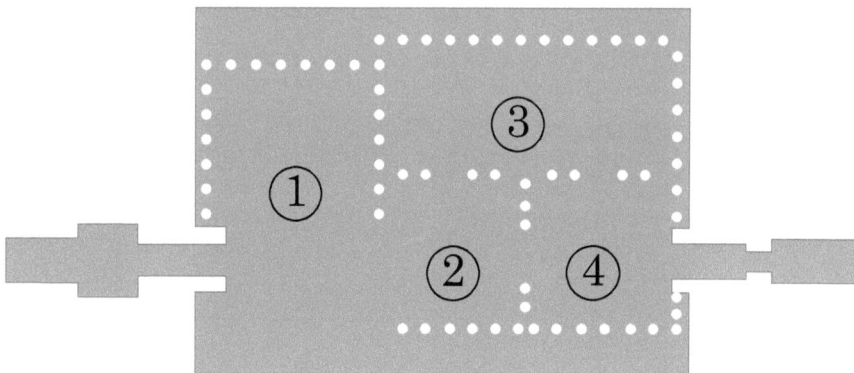

Fig. (14). The geometric structure of the SIW cross-coupled filter.

Fig. (14) shows the geometric topology of the designed filter. The filter contains 4 SIW cavities. The first cavity is larger in size and will be used as its main mode, while the other 3 cavities are smaller in size and used as its main mode.

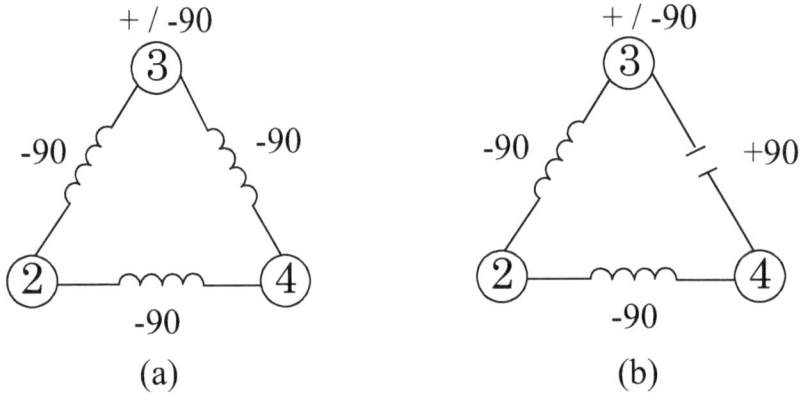

Fig. (15). Phase diagram of the multi-channel coupling path of the CT unit: **(a)** at the resonant frequency of cavity 3; **(b)** at the resonant frequency of cavity 3.

The remaining three smaller cavities in Fig. **(15)** constitute a "three-pole" structure, and its transmission characteristics can be obtained by phase analysis. The circle in the figure represents the resonant cavity, and the inductance and capacitance respectively represent inductive coupling and capacitive coupling.

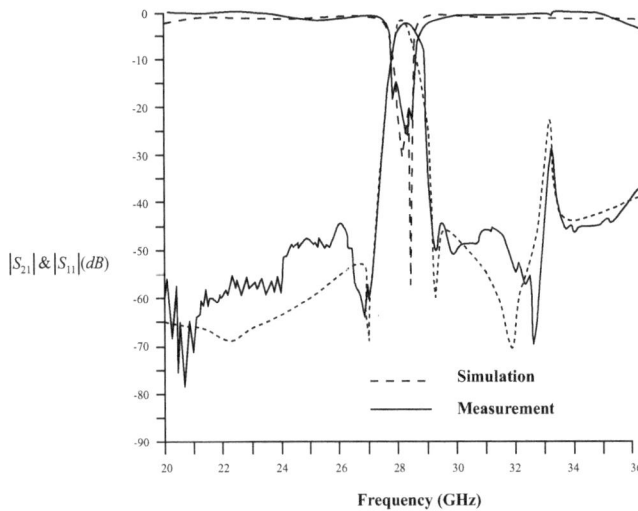

Fig. (16). Filter test and simulation results.

Adjust the size of each cavity and the size of the coupling window to obtain a good passband performance, and adjust the position of the input/output port of cavity 1 so that the transmission zero of the upper sideband is located at 27 GHz. The final result is shown in Fig. (**16**).

ADC/DAC

The mmWave communication will lead to greater bandwidth, which will pose greater challenges to the conversion between analog and digital. The Schreier quality factor of the signal-to-noise distortion ratio is used as a measure of the analog-digital converter,

$$FoM = SNDR + 10\,log_{10}(\frac{f_s/_2}{P}) \tag{10}$$

where the unit of SNDR is dB, the unit of power consumption P is W, and the unit of Nyquist sampling frequency f_s is Hz. The research results show the relationship between the Schreier quality factor of a large number of commercial ADC and the corresponding Nyquist sampling frequency (for most ADC, it is twice the bandwidth) in Fig. (**17**). The dotted line in the figure indicates the envelope of FoM, which is basically constant at 180 dB below the sampling frequency of 100MHz. For a constant quality factor, every 3 dB increase in SNDR or a doubling of bandwidth will cause the power consumption to double. For sampling frequencies above 100 MHz, there will be an additional 10 dB/decade loss, which means that the bandwidth is doubled and the power consumption is 4 times the original.

Although with the development of integrated circuit technology, the high-frequency ADC quality factor envelope in the future will slowly push up. However, ADC with bandwidths in the GHz range still cannot avoid the problem of low power efficiency. The large bandwidth introduced by the NR mmWave and the antenna array configuration will introduce large ADC power consumption. Therefore, both the base station and the terminal need to consider how to reduce the SNDR requirements.

DAC is simpler than ADC under the same accuracy and speed requirements. Moreover, ADC generally introduces cyclic processing while DAC does not. Therefore, DAC is less concerned in the research field, although DAC structure and ADC are very different, DAC can also be described by the quality factor. Large bandwidth and unnecessarily demanding SNDR requirements for the transmitter will result in higher DAC power consumption.

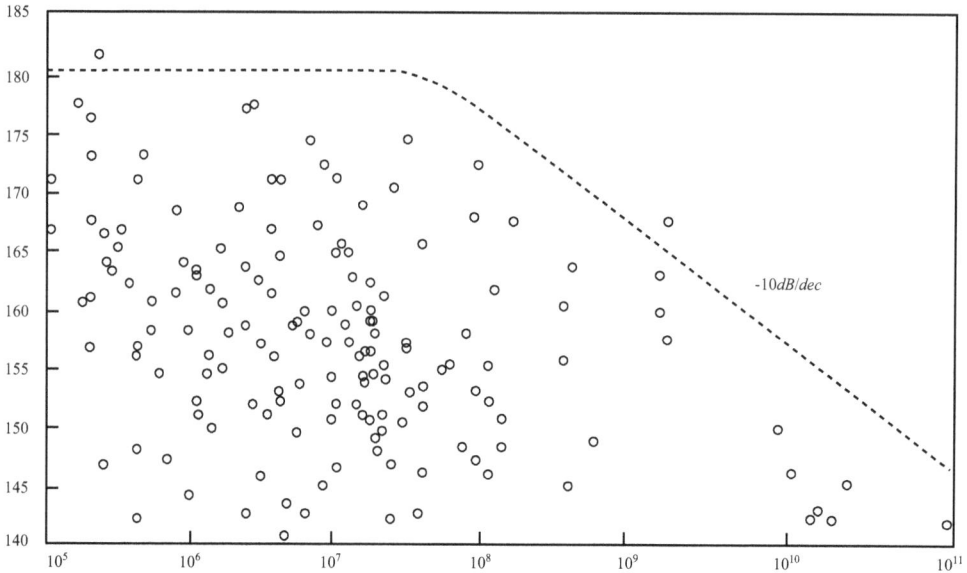

Fig. (17). Schreier quality factor of ADC.

Oscillator

The oscillator is a key component in the mmWave system. It generates signals and provides LO for the transceiver. Traditionally, compound semiconductor technologies such as GaAs or InP can be used to implement mmWave and terahertz frequency oscillators. With the rapid development of silicon-based technology, the frequency of silicon-based oscillators has rapidly increased, making it possible to design low-cost mmWave and terahertz oscillators [20].

In the design of the mmWave CMOS oscillator, a cross-coupled transistor is used as the oscillation core, and a push-pull structure is used to obtain a higher output frequency. Fig. (**18**) shows the commonly used circuit topology of these oscillators. The cross-coupling structure is used as the oscillating core, and a push-pull structure is designed at the output of the oscillator to double the output frequency. Using the push-pull structure, the output frequency can be adjusted by the adjustable capacitor on the resonant network.

Taking into account the connection loss in the measurement, the measured oscillation frequency is about 160.9 GHz, and the output power is about -20 dBm. Taking into account all the mixing and connection losses in the measurement, the measured oscillation frequency of the Y-band oscillator is about 214 GHz, and the output power is about -20 dBm. Figs. (**19** and **20**) show the measured spectral performance results.

Fig. (18). Circuit topology diagram of oscillator.

Fig. (19). D-band CMOS oscillator test results.

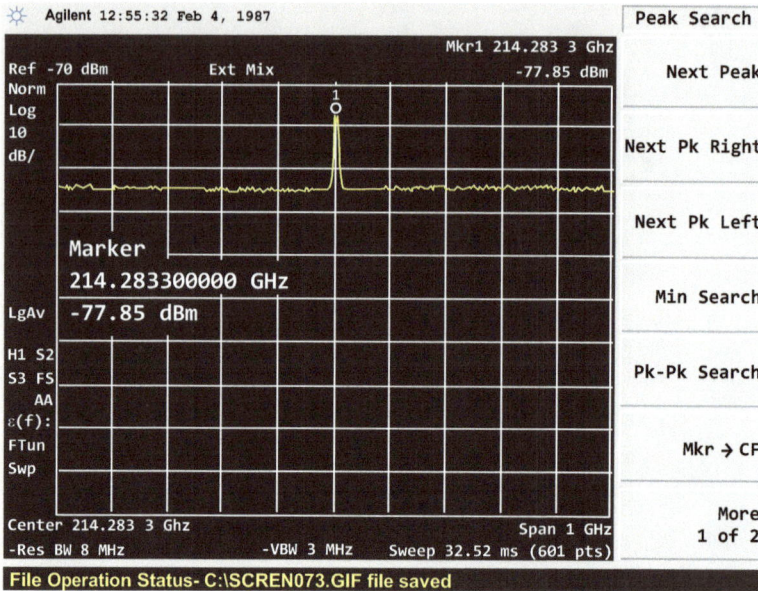

Fig. (20). Measurement results of Y-band CMOS oscillator.

Power Amplifier

In response to the first high-power power amplifier (PA) demand for base stations, traditional CMOS process power amplifiers cannot provide high enough output power, and power amplifiers based on third-generation semiconductor processes such as GaAs (gallium arsenide) and GaN (gallium nitride) can support higher transmission power and larger modulation bandwidth in the mmWave frequency band, which is favored by the industry. There are two main categories when testing the RF parameters of products. The first category is the traditional device parameters for the PA itself, including output power, efficiency, noise figure, S-parameters, *etc*. The second category is the vector error EVM and the adjacent channel leakage ratio ACLR required for 5G broadband modulation signals according to the wireless communication system standards.

The power amplifier used in 33 GHz -37 GHz frequency modulated continuous wave radar does not have high linearity requirements, because the radio frequency signal in the radar system does not contain the information which needs to be transmitted, and its purpose is to transmit a sufficiently high output power. After detecting a far object, the transmitted wave can be reflected back. The saturated output power and efficiency of the power amplifier should be taken as the main design indicators of this design.

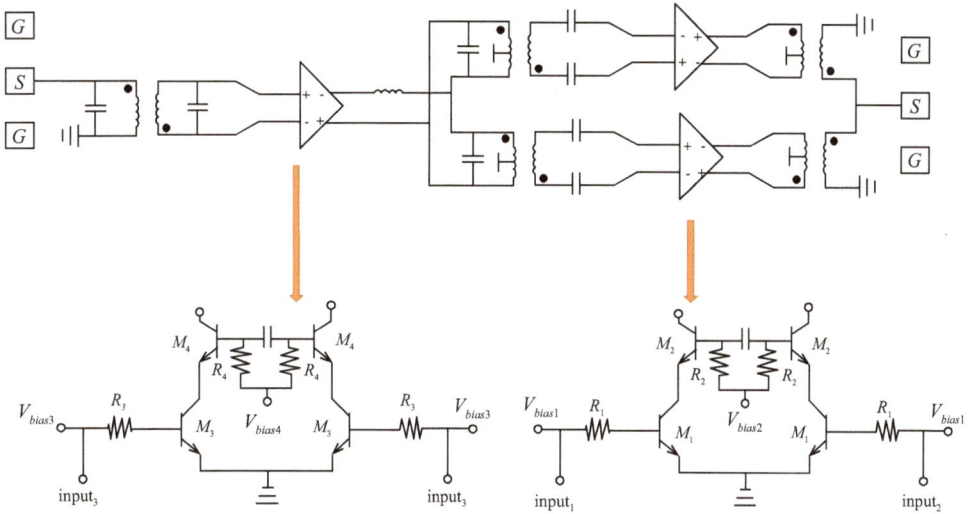

Fig. (21). Power amplifier overall circuit diagram.

Fig. (21) shows the overall circuit diagram of the power amplifier, the overall circuit adopts a differential structure. Due to the needs of the test, the input and output ports need to be balun to convert the differential port impedance into single-ended 50 Ω. In order to achieve a sufficiently higher power gain, the input stage and output stage of the circuit are both cascode structures. Two differential circuits are used in the output stage and power synthesis is performed to increase the output power. The inter-stage matching network between the input stage and the output stage to performs power distribution while achieving impedance matching. The physical separation of the transformer primary and secondary coils is easy to realize the power supply network, which is easy to realize the function of broadband matching. Thus, the input matching, inter-stage matching and output matching network are all realized based on the transformer. Due to the poor stability of the cascode structure of the HBT tube, not only the series resistance of the base of the common emitter is used to improve the stability, but also the series capacitor between the bases of the cascode differential common base transistor is used to destroy the ideal AC ground characteristics between the bases. The base bias voltage and current of the common emitter are connected from the port V_{bias1} through the bias circuit to improve stability, and the base of the common emitter is connected to the supply voltage through a large resistance R_2.

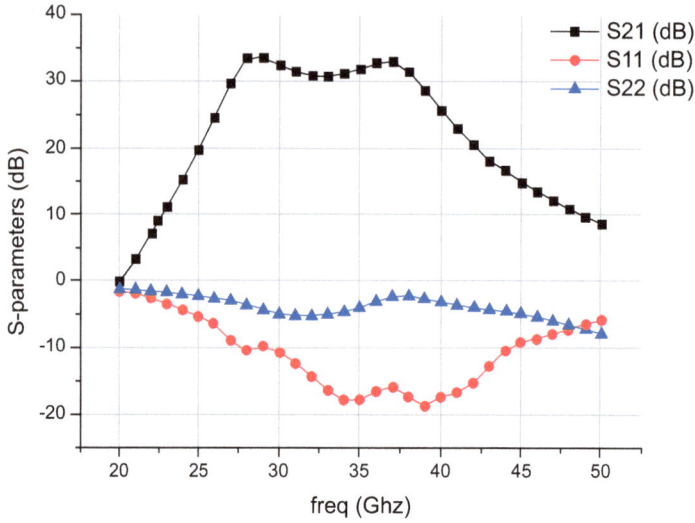

Fig. (22). Power amplifier small signal parameters.

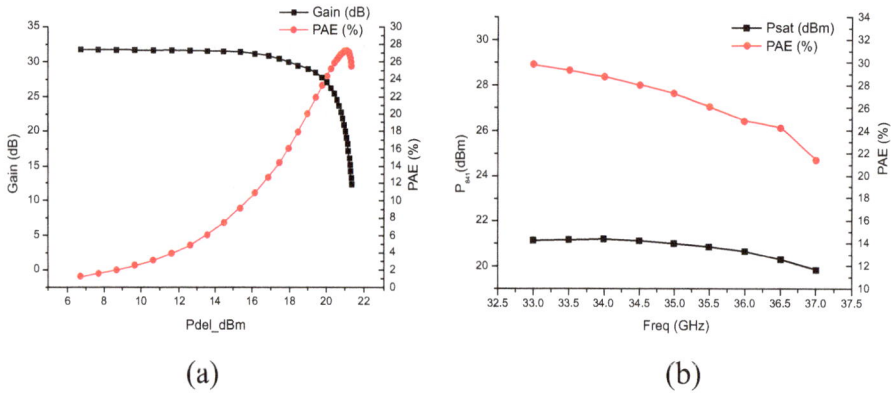

(a)

(b)

Fig. (23). Power amplifier large signal performance(a) 35GHz gain and PAE curve graph (b) 35GHz saturated output power and PAE curve graph.

The small signal parameters of the power amplifier are shown in Fig. (**22**). The 3 dB bandwidth is 27.5 GHz to 38.5 GHz, the in-band power gain S_{21} is greater than 30 dB, the relative bandwidth is 33%, and the in-band $S_{11} < -10$ dB. Fig. (**23**) shows the large-signal simulation performance. Fig. (**23a**) shows the curve of gain and PAE with output power at 35 GHz. The saturated output power is 21 dBm and the maximum PAE is 27.3%. Fig. (**23b**) shows the curves of saturated output power and maximum PAE at different operating frequency points. It can be found that within the operating frequency range of 33~37GHz, the saturated output power is 20 dBm or above, and the maximum PAE is over 21%.

SUMMARY

Firstly, the characteristics of millimeter waves are described. Among them, bandwidth and capacity are the biggest advantages of millimeter waves, which determine the main work range, and the shortwave characteristics of millimeter waves determine the anti-interference characteristics. In view of the characteristics of millimeter waves, the use of millimeter wave offset in communication systems is an excellent choice. At the same time, millimeter waves have a wide range of military uses, such as millimeter wave guided weapons, millimeter wave pulse weapons and military radars, *etc.*

For millimeter waves, the radio wave propagation model needs to be analyzed firstly, which is divided into a large-scale propagation model and a small-scale propagation model for the radio wave propagation model. The large-scale propagation model is designed to describe the change of the received signal power over a long distance between the transmitting end and the receiving end with the distance between the receiving and receiving ends, while the small-scale propagation model is designed to characterize the rapid change of signal strength of electromagnetic waves in a small range Scale fading characteristics. Then the millimeter wave channel model is analyzed. The channel model is the basis of wireless communication system design and communication algorithm technology evaluation. Therefore, it is very important to model the channel accurately. At present, most of the millimeter wave channel modeling uses the time cluster and space lobe method, and the channel modeling method based on ray tracing technology. For the millimeter wave channel model, three international standards have been formulated, namely IEEE 802.15.3c, IEEE 802.11ad and ECMA-387.

Communication carriers above 6GHz can be divided into multi-carrier waveforms and single-carrier waveforms, of which the main performance indicators include spectral efficiency, PARR, time diversity and frequency diversity. Then the massive MIMO antenna technology was discussed, and the basic architecture of the millimeter wave Massive MIMO, the current research direction and the beamforming characteristics of MIMO were also discussed.

For millimeter wave networks in mobile communications, network backhaul technology is very important. There is a brief discussion on 5G NR technology, self-backhaul technology and small base station backhaul technology in the backhaul technology. As a key technology in millimeter wave network backhaul, effective spatial reuse, efficient path establishment, reasonable resource allocation, and convenient access network caching all play an important role.

For the increase of mobile network data traffic, ultra-dense networking is indispensable, and some of the technologies in the networking are particularly important, such as virtualization layer technology, multi-connection technology, anti-interference management, and wireless backhaul methods. However, in such a dense scenario, networking also faces many problems, such as system interference, increased mobile signaling load, cost and energy consumption, and low-power base stations. The network deployment of the network breaks the traditional flat single-layer network structure and builds a multi-layer heterogeneous network. The deployment plan and deployment architecture of the network are discussed.

CONSENT FOR PUBLICATION

Not applicable.

CONFLICT OF INTEREST

The author declares no conflict of interest, financial or otherwise.

ACKNOWLEDGEMENTS

Declared none.

REFERENCES

[1]   J. Lu, D. Steinbach, and P. Cabrol, "Modeling the impact of human blockers in millimeter wave radio links", *ZTE Commun. Mag,* vol. 10, no. 4, pp. 23-28, 2012.

[2]   H.N. Chu, and T. Ma, "An extended 4×4 Butler Matrix with enhanced beam controllability and widened spatial coverage", *IEEE Trans. Microw. Theory Tech.,* vol. 66, no. 3, pp. 1301-1311, 2018.
[http://dx.doi.org/10.1109/TMTT.2017.2772815]

[3]   J. Xu, R. Peng, and C. Zhu, "A study of flames with millimeter-wave radiation", *Defence Technology,* vol. 16, no. 4, pp. 876-882, 2020.
[http://dx.doi.org/10.1016/j.dt.2019.11.004]

[4]   Y. Cao, K. Chin, W. Che, W. Yang, and E.S. Li, "A compact 38 GHz multibeam antenna array with multifolded Butler Matrix for 5G applications", *IEEE Antennas Wirel. Propag. Lett.,* vol. 16, pp. 2996-2999, 2017.
[http://dx.doi.org/10.1109/LAWP.2017.2757045]

[5]   Y. Ma, L. Yang, and X. Zheng, "A geometry-based non-stationary MIMO channel model for vehicular communications", *China Commun.,* pp. 30-38, 2018.
[http://dx.doi.org/10.1109/CC.2018.8424580]

[6]   M.R. Akdeniz, "Millimeter wave channel modeling and cellular capacity evaluation", *IEEE J. Sel. Areas Comm.,* vol. 32, no. 6, pp. 1164-1179, 2014.
[http://dx.doi.org/10.1109/JSAC.2014.2328154]

[7]   G. Sun, and H. Wong, "A Planar Millimeter-Wave Antenna Array With a Pillbox-Distributed Network", *IEEE Trans. Antenn. Propag.,* vol. 68, no. 5, pp. 3664-3672, 2020.
        [http://dx.doi.org/10.1109/TAP.2020.2963931]

[8]   A.A. Zaidi, *"Evaluation of Waveforms for Mobile Radio Communications above 6 GHz", in 2016 IEEE Globecom Workshops.* GC Wkshps, 2016, pp. 1-6.

[9]   C. Park, and T.S. Rappaport, "Short-Range Wireless Communications for Next-Generation Networks: UWB, 60 GHz Millimeter-Wave WPAN, And ZigBee", *IEEE Wirel. Commun.,* vol. 14, no. 4, pp. 70-78, 2007.
        [http://dx.doi.org/10.1109/MWC.2007.4300986]

[10]   C. Kourogiorgas, S. Sagkriotis, and A.D. Panagopoulos, "Coverage and outage capacity evaluation in 5G millimeter wave cellular systems: impact of rain attenuation", *9th European Conference on Antennas and Propagation (EuCAP),* 2015pp. 1-5

[11]   B. Yang, Z. Yu, Y. Dong, J. Zhou, and W. Hong, "Compact Tapered Slot Antenna Array for 5G Millimeter-Wave Massive MIMO Systems", *IEEE Trans. Antenn. Propag.,* vol. 65, no. 12, pp. 6721-6727, 2017.
        [http://dx.doi.org/10.1109/TAP.2017.2700891]

[12]   C. Yu, "Full-Angle Digital Predistortion of 5G Millimeter-Wave Massive MIMO Transmitters", *IEEE Trans. Microw. Theory Tech.,* vol. 67, no. 7, pp. 2847-2860, 2019.
        [http://dx.doi.org/10.1109/TMTT.2019.2918450]

[13]   R.W. Heath, N. González-Prelcic, S. Rangan, W. Roh, and A.M. Sayeed, "An Overview of Signal Processing Techniques for Millimeter Wave MIMO Systems", *IEEE J. Sel. Top. Signal Process.,* vol. 10, no. 3, pp. 436-453, 2016.
        [http://dx.doi.org/10.1109/JSTSP.2016.2523924]

[14]   S.A. Busari, K.M.S. Huq, S. Mumtaz, L. Dai, and J. Rodriguez, *Millimeter-Wave Massive MIMO Communication for Future Wireless Systems: A Survey,* 2018.
        [http://dx.doi.org/10.1109/COMST.2017.2787460]

[15]   J. Góra, and K. Safjan, "Design and optimization aspects of wireless backhaul operated in the millimeter wave spectrum", *25th Annual International Symposium on Personal, Indoor, and Mobile Radio Communication (PIMRC),* 2014pp. 1575-1579
        [http://dx.doi.org/10.1109/PIMRC.2014.7136419]

[16]   M. Jaber, M.A. Imran, R. Tafazolli, and A. Tukmanov, "5G Backhaul Challenges and Emerging Research Directions: A Survey", *IEEE Access,* vol. 4, pp. 1743-1766, 2016.
        [http://dx.doi.org/10.1109/ACCESS.2016.2556011]

[17]   Z. Ning, Q. Song, L. Guo, and A. Jamalipour, "Throughput improvement by network coding and spatial reuse in wireless mesh networks", *Global Communications Conference (GLOBECOM),* 2013pp. 4572-4577

[18]   P.K. Agyapong, M. Iwamura, D. Staehle, W. Kiess, and A. Benjebbour, "Design considerations for a 5G network architecture", *IEEE Commun. Mag.,* vol. 52, no. 11, pp. 65-75, 2014.
        [http://dx.doi.org/10.1109/MCOM.2014.6957145]

[19]   Z. Chen, W. Hong, and Y.U. Chen, "Q-Band transceiver makes MIMO links J", *Microw. RF,* vol. 52, pp. 70-116, 2013.

[20]   J. Chen, W. Liang, P. Yan, D. Hou, Z. Chen, and W. Hong, "Design of silicon based millimeter wave oscillators", *International Conference on Microwave and Millimeter Wave Technology (ICMMT),* 2016pp. 261-263

CHAPTER 3

Antenna Evolution for Massive MIMO

Min Wang[1,*] and **Brian D. Gerardot[2]**

[1] *College of Electronic Engineering, Chongqing University of Posts and Telecommunications, Chongqing, China*

[2] *Institute of Photonics and Quantum Sciences, SUPA, Heriot-Watt University, Edinburgh, EH14 4AS, UK*

Abstract: Massive MIMO technology is one of the key technologies of the 6G communication system and the basis of high-speed transmission in the wireless communication network. At the same time, massive MIMO puts forward new requirements for antenna devices in the communication system, such as active integration, miniaturization, broadband, *etc*. This chapter first reviews the development process and the latest progress of antenna technology in wireless communication. The requirement of a 6G communication network for the base station antenna and the terminal antenna is emphasized. The theoretical basis of massive MIMO technology in the 6G communication system is described in detail, and RIS technology which is closely related to massive MIMO, is introduced. The technical characteristics and development trends of massive MIMO antennas are discussed in detail, including antenna design and synthesis, feed network, and antenna selection technology. Finally, combined with the development of current antenna measurement and calibration technology, the measurement engineering technology closely related to the antenna feeder industry in 6G communication is introduced.

Keywords: Antenna array, Antenna measurements, Beam forming, Feed network, Massive MIMO.

INTRODUCTION

With the further development of the multi-antenna technology theory and the progress of baseband processing ability, RF (radio frequency), and antenna technology, the standardization development of multi-antenna technology is gradually moving towards the direction of further improving the multi-antenna dimension, supporting more users and more parallel transmission of the data stream. In the 6G system, massive MIMO technology, supporting tens, hundreds, and thousands of antennas, will become an important technical way to further

* **Corresponding author Min Wang:** College of Electronic Engineering, Chongqing University of Posts and Telecommunications, Chongqing, China; Tel: 0086 23 62460016; Fax: 0086 23 62468309; E-mail: wangm@cqupt.edu.cn

Xianzhong Xie, Bo Rong, Michel Kadoch (Eds.)

improve the efficiency of the wireless access system, which can satisfy the explosive increase of the user quality and business volume.

The deployment of Massive MIMO can be divided into a distributed structure and a centralized structure. The antenna spacing of the distributed structure is far greater than 10 times the wavelength. In the hot spot area or indoor environment, multiple antennas are distributed in different geographical locations to form different access points [1]. Many access points can be gathered to baseband processing nodes or computing centers through optical fiber or other forms of back-propagation network. In order to achieve high-speed transmission and capacity improvement, the distributed large-scale antenna array is used based on the cooperation between antenna ports. For the centralized large-scale antenna array, the deployment model of small spacing is adopted (small spacing refers to 1/2 wavelength of the electromagnetic wave). By utilization of the characteristics of centralized massive MIMO antenna array with small antenna spacing and the strong correlation between antennas, high gain narrow thin beam with higher spatial resolution can be formed to achieve more functions, such as making space division multiple access with good efficiency, improving the received signal quality and reducing the interference between users greatly, and enhance the system capacity and the spectrum efficiency. The centralized large-scale antenna is also known as large-scale antenna beam forming technology or large-scale antenna due to the use of beam forming signal transmission. Based on the beam forming technology, the centralized small spacing large antenna array plays an important role in promoting the efficiency of frequency band utilization, improving coverage, and suppressing interference. And the centralized large-scale antenna is the most popular technology to design and standardize the large-scale antenna system.

The large-scale antenna beam forming technology plays an important role in different frequency bands. In the Sub-6 GHz frequency band, the large-scale antenna beam forming technology can realize the spatial differentiation of users and suppress the interference effectively through high gain narrow thin beam with higher spatial resolution. In the frequency band above 6 GHz, a two-stage shaping structure with mixed digital and analog signals is generally adopted due to the equipment cost, power consumption, and complexity. The digital phase shifter is used to roughly match the spatial characteristics of signals in the analog domain to overcome the path loss. Then, the user level and frequency selective digital beam forming technique is used to precisely match the channel characteristics in the lower dimensional digital domain. The transmission quality is improved and the interference is effectively suppressed finally. In this case, beam forming technology will play a more important role in making up for the imperfect propagation environment and ensuring system coverage.

The centralized Massive MIMO (referred to as Massive MIMO) and large-scale antenna beam forming technology (referred to as large-scale antenna), based on the small antenna spacing array, are analyzed in the following sections of this chapter., which can form high-resolution, high-gain narrow and thin beams.

OVERVIEW OF ANTENNA IN WIRELESS COMMUNICATION

Evolution of Base Station Antenna

The antenna is a kind of converter to radiate and receive electromagnetic waves. It can be used as a transmitting device to convert high-frequency current into radio waves of the same frequency and can also be used as a receiving device to receive and convert radio waves into the high-frequency current of the same frequency. The antenna is widely used in mobile communication, broadcasting, radio, remote sensing, and other fields. For the mobile communication system, antenna is the converter of equipment circuit signal and electromagnetic wave signal. Since antenna is the entrance and exit of information, its performance affects the performance of the whole mobile network.

With the development of mobile communication systems, the research of base station antenna has entered broadband and multiple frequency era. On the one hand, the evolution of mobile communication systems is a step-by-step process, and the coexistence of 2G, 3G, and 4G systems will be maintained for quite a long time. Multi-system common station and multi-system common antenna are economical and effective solutions. On the other hand, it is urgent to develop a compact and wide band base station antenna with increasing attention to visual pollution and electromagnetic radiation pollution.

Since the 1980s, the development of mobile communication technology has comprehensively promoted the evolution of base station antenna technology. The early base station antenna is omni-directional, which requires four antenna elements arranged around the vertical axis to obtain the omni-directional radiation pattern. The sector division of coverage cell makes the base station antenna develop into directional antenna with the popularity of cellular mobile communication system. Due to the expansion of channel capacity, compatibility of operation system and the flexibility of service mode, the working frequency band of base station antenna is prominently extended in recent years. Therefore, broadband, multiple frequency, miniaturization and integrated base station antenna which can meet the requirements of various systems are the research hotspots of base station antenna [2]. Based on the long-term research, the development trend of the mainstream base station antenna are as follows.

Wide Band Antenna

The frequency range, in which the antenna performance index such as beam width, impedance match, isolation and other characteristics meet the system requirements, is generally called the bandwidth of the antenna, also known as the operation bandwidth of the antenna. In particular, the bandwidth of base station antenna is described by standing wave ratio bandwidth and beam width bandwidth. Wide band is a relative measure of antenna, which varies according to different applications. f_{max} and f_{min} are the upper limit and lower limit of the operation bandwidth, and f_c is the central frequency. The relative operation bandwidth is defined as:

$$B_p = \frac{f_{max} - f_{min}}{f_c} \tag{1}$$

Generally speaking, there is no strict definition of the broadband antenna, which depends on the special antenna. If the relative bandwidth B_p is greater than 20%, it is often regarded as broadband antenna. In particular, the antenna which can cover two or more different standards simultaneously is called broadband antenna for the base station antenna of mobile communication. According to wideband antenna technology, the main methods include tapered structure, application of parasitic elements, modification of Quality Factor (Q Value) and frequency-independent antenna.

Although the research on the miniaturization and broadband of base station antennas has been very abundant, broadband miniaturized base station antennas which can be applied in engineer solution are rarely to be found. Due to the special application background of base station antennas, both performance and cost of the base station antenna must be considered. Dual polarization base station antenna is widely used in outdoor base station antenna to obtain polarization diversity gain, to which the broadband technology and miniaturization technology could hardly be applied. On the other hand, complex microstrip antenna structure has not been widely used in base station antenna for cost factor.

Miniaturization Technology for Base Station Antenna

The miniaturization of base station antennas refers to reducing the size of the antenna in the fixed frequency range of work. The smallest possible antenna structure is used to improve the radiation performance of the antenna and facilitate integration of system and reduce costs. However, the various parameters of the antenna influence each other, so that the miniaturization of the antenna requires the sacrifice of other indicators. Bandwidth *BW* is closely related to the quality factor *Q*, which is usually defined as:

$$Q = f_c / BW \tag{2}$$

where *BW* is the difference between f_{max} and f_{min}, which represents the absolute bandwidth of the antenna. The quality factor of the antenna is inversely proportional to the bandwidth. If the size of the antenna decreases, the radiation impedance will decrease, and the quality factor *Q* will increase accordingly. In this way, it can be found from the Equation (2) that the reduction in antenna size and the increase in the working bandwidth of the antenna are mutually restrictive, and miniaturization will cause the bandwidth to narrow.

As shown in Fig. **(1)** , the basic condition of the quality factor Q of the linear polarization antenna has been given for a spherical space

$$Q_{min} = \frac{1}{k^3 a^3} + \frac{1}{ka} \tag{3}$$

where *k* is the wave number. According to Equation (3), the spherical space surrounding the antenna needs to be more fully utilized to obtain a lower antenna value in the actual antenna design process. This theory is the basis to realize the complementarity between broadband and miniaturization.

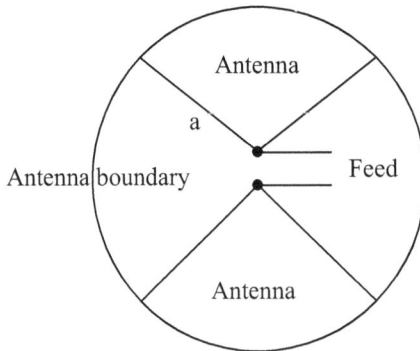

Fig. (1). Radiation schematic diagram of miniaturized antenna.

The design of general base station antenna often starts from the optimization of antenna indicators, and the size of the base station antenna is mostly calculated according to theory. Therefore, the size of the base station antenna is completely determined by the operating frequency, which becomes monotonous and heavy. Furthermore, it is difficult to meet the various requirements of the increasingly complex application environment on the base station antenna. Thus, broadband miniaturized base station antennas have three advantages compared with conventional base station antennas.

Broadband miniaturized base station antennas can reduce the size and weight of the antenna, which not only reduces the cost of antenna processing raw materials, but also provide a prerequisite for the RRU (Radio Remote Unit) tower. On the other hand, the weight of the miniaturized antenna is reduced.

Broadband miniaturized base station antennas can be better used in multi-system antennas, which is the key technology used in the construction of co-site co-location network to replace the chronic problem of the large size of conventional antennas. The application of broadband miniaturized base station antennas solves this problem to a certain extent. On the other hand, broadband miniaturized base station antennas can reduce wind resistance, and be simplified installed.

The broadband miniaturized base station antenna is an important part of the broadband miniaturized base station system. With the rapid development of wireless communication technology, basic network construction is developed in the direction of supporting faster data services. Research on broadband miniaturized base station antennas has gradually become the focus of attention. Due to the special application environment, the broadband miniaturized base station system is required to have the characteristics of flexible form, miniaturization and easy beautification. Since the antenna part has been unable to break through the limitation of space size, the utilization of broadband miniaturized base station antennas will enable the rapid development of broadband miniaturized base station systems.

Active Base Station Antenna System

The emergence of integrated active antennas (AAU) is in line with mobile broadband requirements, as shown in Fig. (**2**) . The integration of radio frequency modules and antennas not only simplifies site deployment, but also reduces system feeder loss, which improves network coverage performance by more than 10%. One AAU can meet the requirements of multi-band site coverage for one sector based on the fusion design with multi-frequency antennas, which efficiently improves site environmental friendliness, and reduce site load. It has been proved that the integrated AAU technology can not only meet the increasingly stringent deployment requirements, but also can promote more new antenna array technologies and new features in the future, as shown in Fig. (**3**).

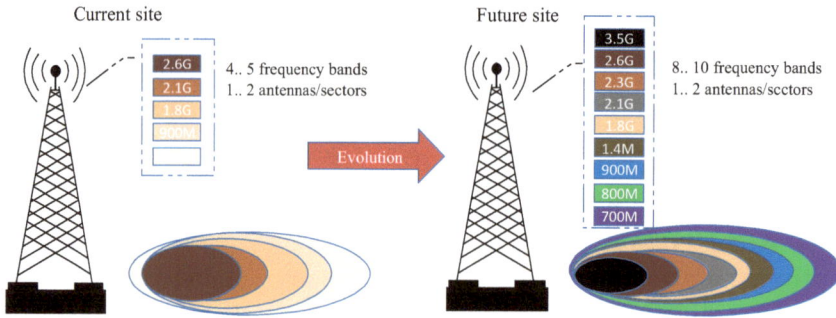

Fig. (2). Evolution of integrated antenna for future base station site.

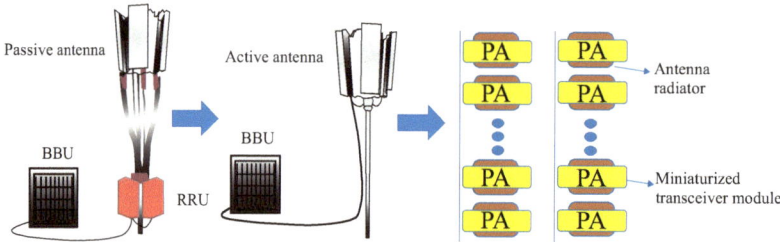

Fig. (3). Evolution from traditional passive base station antenna to Massive MIMO antenna.

The antenna and RRU of the traditional antenna feed system are separated from each other, and the interface is a standardized interface. Their respective performance can be verified through independent tests [3]. However, the integrated active antenna is the integration of the antenna and the radio unit, and the connection between them is a non-standard interface. The RF index performance and radiation performance of the active antenna cannot be fully reflected in a separate manner. Integrated test methods need to be considered, that is, some conduction indicators need to be considered for air interface testing.

While the air interface test is based on the field test, the traditional antenna test is to use a single tone signal to test the amplitude and phase of the corresponding field. However, for the integrated active antenna, the signal may work in different standards, different modulation methods and different bandwidths in its working mode. The field-based test system needs to support amplitude and phase tests of service signals in various configuration scenarios.

For the test of the integrated active antenna pattern, how to select the test signal of the pattern is also a new problem faced by the integrated test, which requires research, analysis and discussion of definitions.

Multiple Beam Antenna System

Due to the rapid development of millimeter wave communication technology, the research of millimeter wave antenna has attracted interest of many scholars. The massive MIMO system also needs the multiple beam antenna to transmit orthogonal channels. The passive beamforming network becomes one the research focuses in multiple beam antenna design. Multi-beam antenna technology uses the sector splitting idea to split the traditional 120 ° sector into N (N≥2) sub-sectors, and correspondingly use N sub-beams to cover each sub-sector in depth. The coverage diagram is shown in Figs. (**4** and **5**). Network capacity can be increased to N times the original through multi-beam antenna technology. The depth of the coverage area can be enhanced by the high gain of the multi-beam antenna, and each beam can be finely directed. This enables the use of multi-beam antenna technology to effectively solve the problems of insufficient system capacity and insufficient coverage depth while maintaining the existing site resources and spectrum resources, allowing the existing network to play a better role.

Since the directional radiation pattern is regularly formed with antenna array, the feed network is significant for the performance of beams [4]. Considering the performance and cost of antenna system, the implementation of analog beam forming network (BFN) is generally used in the field of base station antennas. The Butler matrix is one of the most classic feed networks for the multiple beam antennas. When excitation is applied to any input port of the Butler matrix network, a specific amplitude-phase distribution relationship can be formed between each output port. Using this feature of the Butler matrix network to beam-form the antenna can make the antenna radiate multiple beams directed at demand to realize the multi-beam function of the antenna.

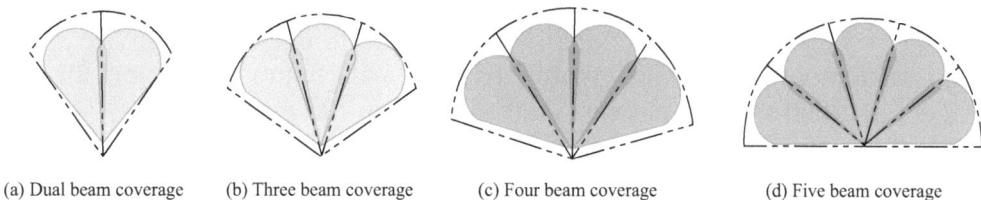

(a) Dual beam coverage   (b) Three beam coverage   (c) Four beam coverage   (d) Five beam coverage

Fig. (4). Extended coverage with multiple beam antenna.

Split antenna can increase capacity to 1.7-2.2 times

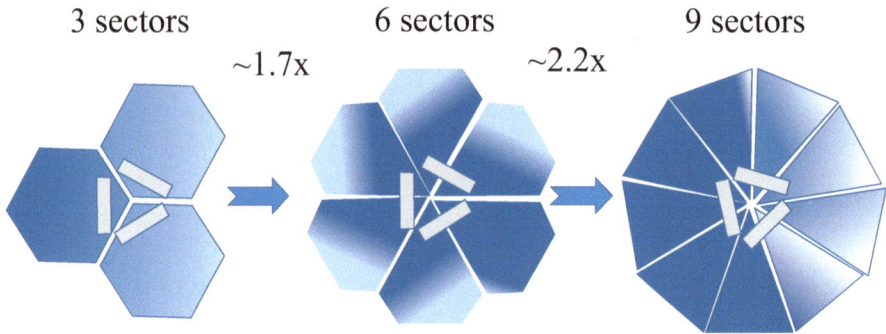

3 sectors   6 sectors   9 sectors

~1.7x   ~2.2x

Fig. (5). System capacity increased with multiple sectors.

Development of User Equipment Antenna

Due to the rapid development of mobile communication technology, the mobile communication terminal industry has gradually emerged. The mobile phone as a terminal product has gradually become an indispensable item in people's life due to its portability and practicability. The mobile phones were initially called feature phones, which were only used to provide basic communication tools for single voice service. With the advent of the smart era, mobile phones have developed more and more functions to better improve the quality of life of users, such as navigation, FM radio, Bluetooth, near field communication (NFC), and so on. To support multi-functions and reserve space for more and more sensors and other hardware, mobile phone antennas which transmit and receive radio waves need to cover multiple frequency bands and have as little space as possible.

The design and deployment of terminal antennas is a problem in the application of MIMO technology and Massive MIMO technology on the mobile terminal side. The biggest challenges for the mobile terminal multi-antenna design are how to place two or more antennas in a mobile terminal with limited space, and meet the low coupling between the antenna elements to work at the same time. To fulfill the design requirements of smaller signal correlation, the distance between the antenna elements should be at least half of the working wavelength, which is difficult to achieve for mobile terminals with a small form factor. For this reason, two ideas through research experiments are put to solve the problem.

Electromagnetic Decoupling of Antenna Elements

The electromagnetic coupling is mainly formed by the surface wave which is excited by the current on the antenna. Therefore, improving the electromagnetic coupling by increasing the isolation between antenna elements mainly means suppressing the mutual interference of surface waves. The specific methods are placing the antenna orthogonally, adopting the electromagnetic complementary structure antenna, introducing the structure between the antenna elements, and changing the floor structure.

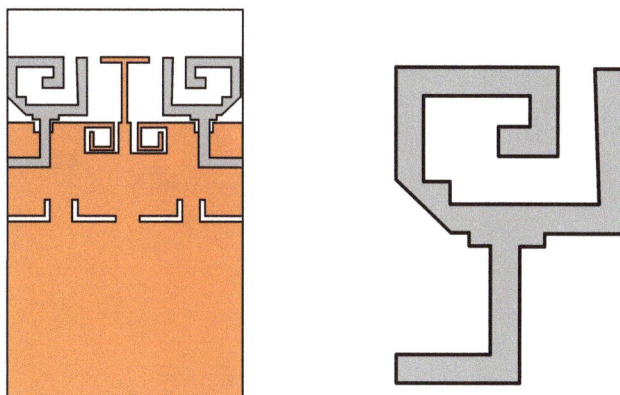

Fig. (6). High isolation terminal MIMO antenna with floor branches.

Orthogonal placement of antennas means that the current paths of the antenna elements are placed orthogonally to significantly reduce electromagnetic coupling and improve isolation. This method can also realize the polarization diversity of the antenna radiation pattern and is mainly suitable for microstrip planar monopole antenna and dipole antenna.

The electromagnetic complementary structure antenna consists of a group of coaxial dipoles and slotted dipoles. The antenna can be placed side by side or vertically, which can be flexibly selected according to the specific application environment and make better use of space. A set of electromagnetic complementary antenna pairs composed of coaxial dipoles and slotted dipoles is used to achieve the antenna isolation of 40 dB.

The introduction of structures between antennas can also improve the isolation between antenna elements. The structure is similar to the working principle of the filter, which can cut off the surface waves between the antennas to achieve high isolation, as shown in Fig. (**6**) and Fig. (**7**). Compared with ordinary metal structure, the antenna isolation using the structure can be increased to more than 40dB.

The defected ground structure (DGS) structure is a special shape pattern which is *etc*hed on the antenna floor, as shown in Fig. (**8**). The DGS structure can use the combined effect of inductance and capacitance to realize the band-stop function to improve the isolation between antenna elements. The common structures include the quarter working wavelength slot, the polygonal resonance structure, and the addition of short-circuit branches.

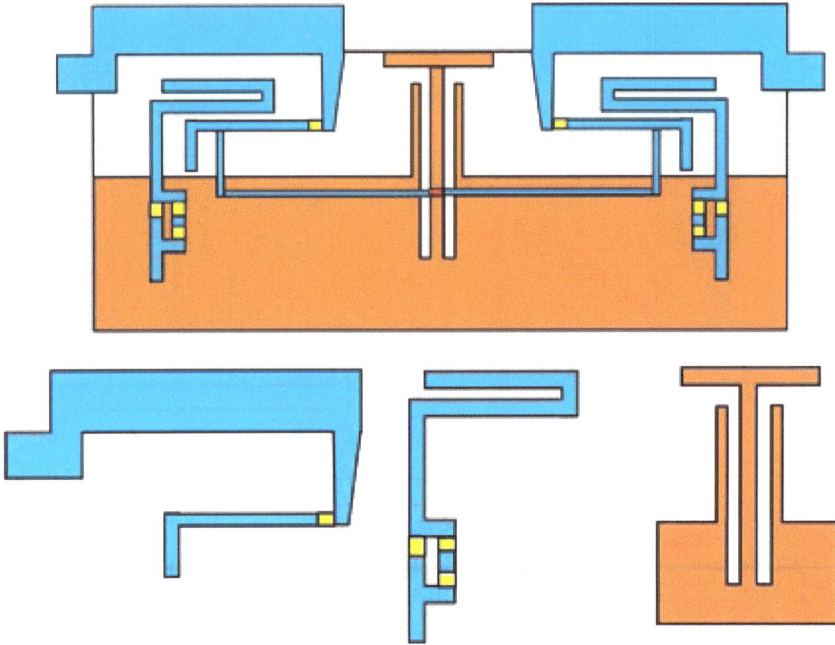

Fig. (7). Terminal MIMO antenna with neutral line.

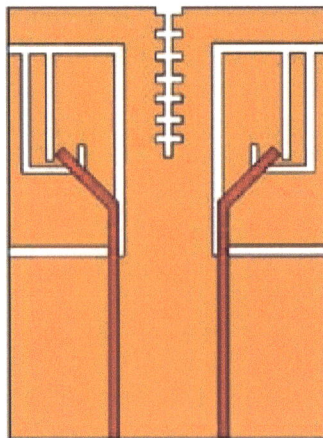

Fig. (8). Defective ground for high isolation terminal MIMO antenna.

Design of Integrated Antenna

The radiation characteristic of the antenna-floor integrated structure antenna is mainly determined by the current which is distributed on the floor. The radiation performance of current distribution is similar to the typical dipole. This method is based on the fact that the radiation characteristics of the mobile terminal antenna in the low-frequency band are mainly determined by the floor waveguide mode, and the antenna element works as a coupling one. The floor waveguide mode is effectively excited by the antenna element as a simple non-resonant unit. Therefore, multi-antenna technology can be realized by placing traditional self-resonant antennas and simple coupling elements in the mobile terminal. The integrated antenna based on floor radiation is mainly divided into three parts,which are mobile terminal floor, coupling element, and matching circuit. The matching circuit placed on the floor circuit board mainly realizes the impedance matching of the antenna, which is very different from the traditional self-resonant antenna that relies on the three-dimensional metal antenna element structure to achieve impedance matching. In addition, the coupling element is required to couple energy to the floor waveguide mode most effectively to achieve radiation. Traditional self-resonant antenna elements need to achieve impedance matching and it is difficult to couple energy to the floor waveguide mode.

Both sides of the printed circuit board in the mobile terminal have continuous grounding layers, and a metal surface is used as an electromagnetic compatibility protective cover. The size and position of the protective cover in different mobile phones are different. However, the ground layer and the metal protective cover in the terminal can be used as the metal ground plate based on the theory of RF technology. In the past two decades, the effect of the grounding plate in the terminal on the performance of the entire antenna structure was studied. The results show that the main waveguide mode excited by the floor currently has a significant impact on the total radiation performance of the floor and the antenna element. Studies have shown that the self-resonant antenna (such as PIFA antenna) in a typical mobile terminal with a working frequency of 900MHz and a relative bandwidth of 10% (S11<=-6dB) only accounts for 10% of the total radiated energy. The remaining energy is radiated by the current which is distributed on the floor. And the radiation pattern of the current is similar to the half-wave symmetric oscillator pattern. Miniaturization of the antenna element can be achieved by using the mobile terminal floor as the main radiator.

Due to the advantages of the metal frame, such as metallic texture, fashion beauty, and greater structural strength, the metal frame design has gradually become the choice of mobile phone antenna manufacturers. However, since the mobile phone antenna is usually placed on the top or bottom of the mobile phone, when the metal frame design is introduced, the performance of the mobile phone antenna will deteriorate. Thus, the miniaturization of multi-band metal frame mobile phone antenna design is meant to cater to the current mobile phone market.

In order to meet the different needs of modern mobile communication, vehicles are loaded with various functional antennas, which can be used in emergency calls, entertainment, navigation and positioning. The performance of automobile antenna is of great significance to improve the quality of service (QoS) in vehicle wireless communication system [5].

The vehicles will be combined with more communication devices to achieve more comprehensive and intelligent driving functions in the future traffic systems. An important challenge is how to integrate the antenna into a specific small volume model, such as the shark-fin shaped roof antenna. Meanwhile, how to realize a wide range of services in several frequency bands is also an crucial problem to solve. While multiple antennas of different frequency bands can be integrated into a fixed volume model, it is significant to reduce the coupling between the antennas and improve the radiation efficiency. Due to the difficulty in improvement of the antenna isolation, the multiband antenna is widely researched to cover multiple frequency bands.

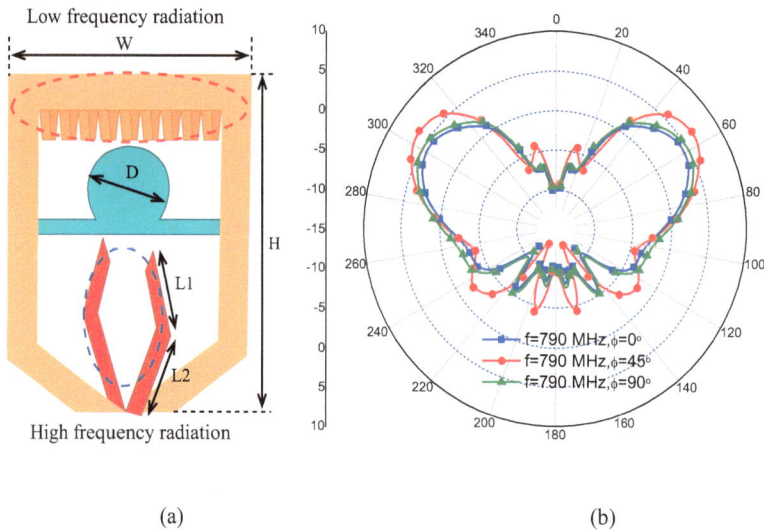

(a)   (b)

Fig. (9). Multiband automobile antenna for V2X communication.

A kind of multiband automobile antenna is proposed in Fig. (**9**), which can cover a wide frequency band for future automobile communication. The impedance matching performance of the antenna is improved by loading the toothed capacitor and the impedance matching disc.

Requirements of Communication Technology on Antenna Technology

The MIMO technology relies more on the spatial freedom brought by the antenna array to show its performance advantages. The existing MIMO transmission scheme which is limited by the traditional base station antenna architecture can only control the signal spatial distribution characteristics in the horizontal plane rather than the vertical dimension in the 3D channel. The feature of MIMO technology for improving the overall efficiency and performance of the mobile communication system and the user experience are not reflected in the existing MIMO technology. With the evolution of antenna design, the development of active antenna system (AAS) technology has a great impact on the underlying design and network structure design of the mobile communication system. It will promote MIMO technology from the traditional optimization design for 2D space to higher dimensional space expansion.

The signal spatial distribution characteristics are adjusted by the base station through beamforming or precoding. The adjustment process can be roughly divided into two levels. The first level is the sector-level adjustment of the common channel and the common physical signal which is adjusting the coverage parameters of the sector according to the network optimization goal, such as sector width, pointing, down tilt, *etc*. The operation of this level can also be referred to as sector-level shaping. The shaping method is not optimized for a small-scale channel of a certain user equipment (UE). The adjustment of sector shaping is a relatively static process. The second level of adjustment is UE-level dynamic shaping or precoding performed for each UE, and its purpose is to match the transmission for each UE with its channel features.

The spatial distribution of signal power can be decomposed into two two-dimensional spaces (the horizontal plane and the vertical plane) in the three-dimensional coordinate system. The physical antenna port of the existing base station antenna structure corresponds to a linear array arranged in a horizontal direction. Adjusting the amplitude of each physical antenna port and the relative phase between the physical antenna ports is equivalent to the distribution of the control signal in the horizontal dimension. The sector shaping or UE-level dynamic shaping can be implemented in the baseband through the antenna mapping module. However, for the array of elements corresponding to each antenna port, since there is no corresponding physical antenna port corresponding to one of them, the weighting coefficient of each element cannot be directly adjusted in the baseband. The flexibility of adjusting the signal power distribution in the vertical dimension is limited to a certain extent.

For sector shaping, the declination angle can be controlled at the radio frequency by adjusting the time delay and attenuation of the radio frequency cable connected to each element, or the elevation angle of the base station antenna panel can be adjusted mechanically. However, for the service transmission of each UE, dynamic optimization for small-scale channels cannot be achieved in the vertical dimension. According to the current passive base station antenna structure, the MIMO transmission method can only optimize the transmission process in the horizontal dimension, but does not fully match the actual three-dimensional channel. The freedom of the signal in the vertical dimension cannot be fully utilized in the transmission. Cell splitting or further sector splitting is also an important means to expand the capacity of the system. Due to the traditional base station antenna structure, it is impossible to achieve vertical sectorizing without increasing the antenna and radio frequency equipment. Multi-faceted antennas are often required to cover the areas with different vertical angles separately, such as different height ranges of high-rise buildings.

Given the shortcomings of the existing base station antenna structure in the vertical dimension, a natural idea is to increase the physical antenna ports in the vertical dimension to realize independent control of each element in the baseband. Based on the existing passive antenna structure, the problem is not the technical principle but the engineering realization. According to the passive antenna structure, the RF circuit and the physical antenna port are connected by RF cables. And the number of physical antenna ports determines the number of RF cables. The number of RF cables is considerable in the case of providing horizontal dimension physical antenna ports. To decrease the loss caused by RF cables and the workload of RF cable installation, and reduce the influence of the weight and wind resistance of the cables on the base station tower, the development trend of the base station structure is to install the RF circuit part (RRH, remote radio head) as close to the antenna as possible. It is not feasible to increase the number of physical antenna ports on a large scale, even if the RRH structure is adopted on the tower where the RF cables are already installed.

For the base station and antenna structure, since the small amount of high-power amplification equipment in the RRH or tower amplifier is replaced by a large number of relatively low-power amplifiers integrated with the array in the AAS array. And there is a large heat dissipation space on the ASS antenna panel for the power amplifier. ASS array can also support higher total transmit power more stably without other auto radiating equipment. The combination of RF module and antenna system further reduces the number of tower equipment and the corresponding rental costs.

The use of vertical-dimensional ports is of great significance for enhancing sector-level shaping and UE-level shaping capabilities. Sector-level shaping focuses more on implementation, while the improvement of UE-level shaping involves corresponding standardization work. In the traditional cellular network, capacity expansion can only be achieved by splitting the sector or setting up a new cell when the traffic of a sector exceeds the carrying capacity. The above method is not only expensive but also has many obstacles in site selection and other issues. Based on the AAS array, the original site and equipment can be utilized fully to re-divide the original sector in the vertical dimension into inner ring and outer ring sub-sectors with different down-tilt angles to achieve vertical sectorizing and further improve spectrum utilization efficiency. For high-rise buildings, multiple areas of different heights can be covered by vertical sectorizing. With the flexibility of adjusting the vertical dimension of the AAS array, it is also possible to achieve independent optimization of sector coverage of multiple air interface modes occupying different carrier frequencies in the same frequency band. The AAS array can also realize the independent optimization of the uplink and the downlink so that the uplink can better cover cell edge users and the downlink can avoid unnecessary interference to neighboring cells.

The emergence of large-scale antenna beamforming technology theory has laid a theoretical foundation for the further expansion of MIMO dimensions. The increasingly mature application conditions of ASS in commercial mobile communication systems have created favorable conditions for the further development of MIMO technology in the direction of large-scale and 3D. Based on the large-scale antenna array, the base station can form a pencil beam with high spatial resolution in the three dimensional space, which can provide flexible spatial diversity capabilities, improve the received signal and suppress the interference to achieve higher system capacity and spectrum utilization efficiency.

Development Trends of Massive MIMO Antenna for 6G

Although the academia has carried out relatively extensive and in-depth research on Massive MIMO beamforming technology, there is still a need for further in-depth research on several key technical issues at the important turning point in the transition from theoretical research to standardization and practicality [6].

The performance of the Massive MIMO technical solution is very closely related to the application scenario and deployment environment. Therefore, it is necessary to combine the deployment scenarios and business requirements of the next-generation mobile communication system to study the applicable scenarios of large-scale multi-antenna technology in a targeted manner. And measure and model channel parameters for its typical application scenarios and channel characteristics. This work will provide directional guidelines for Massive MIMO antenna selection, technical solution design and standard solution formulation. At the same time, modeling based on measured channel parameters for typical application scenarios will also provide an important basis for accurately constructing a technical solution evaluation system and accurately predicting the performance of the technical solution in the actual application environment.

The performance of the MIMO system is very dependent on the form of the antenna array used by the system and the characteristics of the propagation environment. And the same is true for Massive MIMO technology. The spatial resolution of the Massive MIMO channel is significantly increased in the Massive MIMO wireless communication environment when the base station is equipped with a large-scale array antenna on the side especially. Whether there are new features in the channel needs further discussion. There is no correlation and mutual coupling between the antennas in the ideal model. Therefore, adding additional antenna elements will significantly increase the degree of freedom of the system. However, the characteristics of the spatial channel are difficult to be so ideal in actual systems. The distance between the elements in the array is usually small. And there may be insufficient scatterers in the propagation environment. The above factors will affect the spatial freedom of Massive MIMO systems.

It is often necessary to perform operations on a large number of high-dimensional matrices in the transmission, detection and scheduling process of MIMO with the increase of the number and scale of antennas, the increase of the number of users and the improvement of bandwidth, which leads to a significant increase in system complexity. This problem is particularly prominent in high-frequency systems. In addition, it is necessary to consider the balance of overhead, complexity and performance brought about by the increase in the number of antennas in terms of system design such as reference signals, feedback mechanisms, control signaling, broadcast/public signal coverage, access and handover. The problem mainly involves 4 aspects including beamforming technology, channel measurement and feedback technology, coverage enhancement technology and high-speed mobile solutions and multi-user scheduling and resource management technology.

The performance gain of Massive MIMO is mainly guaranteed by the quasi-orthogonal characteristics between multi-user channels formed by a large-scale array in terms of beamforming technology. However, due to the existence of many non-ideal factors in the equipment and the propagation environment, it is still necessary to rely on the design of downlink transmission and uplink reception algorithms to effectively suppress co-channel interference between users and even cells in order to obtain stable multi-user transmission gain in actual channel conditions. And the computational complexity of the transmission and detection algorithm is directly related to the number of antenna arrays and users. In addition, precoding/beamforming algorithms based on Massive MIMO are directly related to array structure design, design cost, power efficiency and system performance. Therefore, the balance between the computational complexity and system performance of the Massive MIMO transmission and detection scheme will be the primary issue for the practical application of this technology.

Since channel state information measurement, feedback and reference signal design technology are critical to the realization of MIMO technology, they have always been the core content of MIMO technology research. The research, evaluation and verification and standardization scheme design for this problem are of extremely important value for the practical development of Massive MIMO technology. The pilot resource is an important and limited resource in large-scale antenna systems. Due to the limited coherence time in a multi-cell scenario, the pilots need to be reused between cells. The pilot pollution caused by pilot reuse is a bottleneck that affects performance in Massive MIMO systems. Therefore, it is necessary to study the sounding, design mechanisms of reference signal and efficient allocation mechanisms of reference signal resource. At the same time, the accuracy of channel estimation is also very important in a pilot-contaminated environment. In order to effectively suppress interference, it is necessary to study more effective channel and interference estimation methods and network assistance mechanisms to enhance system performance. In addition, reference signal design is closely linked to the channel state information feedback mechanism, which will directly affect the efficiency and performance of large-scale antenna systems. The feedback of channel state information is an important foundation for basic functional modules such as resource scheduling, link adaptation, and MIMO in the wireless access system. As mentioned earlier, the accuracy of channel state information has a vital impact on the frequency band utilization efficiency of the entire system for MU-MIMO. The conflict between the improvement of feedback accuracy and the overhead will become more prominent for the codebook-based feedback mechanism as the antenna scale increases. The advantages of the channel reciprocity feedback mechanism based on the unique TDD system in terms of channel state information accuracy, the downlink measurement reference signal overhead and the feedback overhead are also increasingly obvious in this case.

The expansion of antenna scale will bring huge gains to the coverage of traffic channels in terms of coverage enhancement technology and high-speed mobile solutions. But it will bring a lot of negative effects for broadcast channels that need to effectively cover all terminals in the entire cell. In addition, the issue of how to provide reliable high-speed signal transmission also needs to be considered in high-speed mobile scenarios. The biggest challenge for Massive MIMO systems is the severe time-variation of channel information in this scenario. At this time, beam tracking and beam broadening technologies that are less dependent on the acquisition of channel state information can effectively use the array gain of Massive MIMO to improve data transmission reliability and transmission rate, which is worthy of our further exploration.

Massive MIMO provides a better spatial granularity and more spatial freedom for wireless access networks in terms of multi-user scheduling and resource management technology. Therefore, considerable performance gains will be obtained for the multi-user scheduling technology, the service load balancing technology and resource management technology based on Massive MIMO.

Massive MIMO beamforming technology was in connection with the non-cooperative transmission technology for single Macrocell scenarios when Bell Labs proposed the Massive MIMO concept for the above-mentioned scenarios in 2010. According to this concept, the increase in system capacity that large-scale antennas can bring comes from the whitening of spatial channels by large-scale arrays, and the nearly orthogonal state between multi-user channels obtained therefrom, rather than through inter-cell or Cooperative acquisition between transmission points. In other words, after applying a large-scale array, there is no need for collaboration under ideal circumstances. However, if there is some kind of cooperation mechanism between the cells, it is actually possible to further improve the performance of Massive MIMO. For example, the so-called pilot pollution problem can be avoided as much as possible through a cooperative mechanism. In addition, another issue that needs attention is the so-called flashlight effect. After using the beam shaping technology, the neighboring interference received by the user can be compared with the beam formed by a flashlight. With the changes in the neighboring cell scheduling situation or the movement of the scheduled users in the neighboring cell, the beam that causes interference sways like a flashlight or turns on/off intermittently. The resulting interference fluctuations will be very severe. In this case, it is difficult to accurately predict the interference fluctuations through CQI reporting. If there is no coordination or cooperation mechanism, this unpredictable and violently fluctuating interference will have a very serious impact on link adaptation operations such as AMC (Adaptive Modulation and Coding) (especially after the use of large-scale arrays, The array gain is very high and the impact will be more serious). Therefore, in practical applications, collaboration is also necessary for centralized large-scale antenna systems in wide-area coverage scenarios. The collaboration of large-scale antenna systems needs to focus on centralized large-scale antenna collaboration, distributed large-scale collaboration, and heterogeneous scenarios centralized and distributed coexistence.

As mentioned above, the passive antenna structure is generally adopted in the existing mobile communication system. Each antenna port needs an independent RF cable to be connected to it. When the number of antenna ports that need to be independently controlled gradually increases, a large number of RF cables will bring unimaginable difficulties to project realization and subsequent operation and maintenance. The existing antenna system generally can only support independent adjustment of the beam for each user in the horizontal dimension. However, the beam shape can only be uniformly set for sector coverage requirements in the vertical dimension. Therefore, MIMO transmission based on this array is also called 2D-MIMO.

To solve the above problems, RF (part of the baseband function) can be combined with a traditional passive antenna array to form an active antenna technology. The optical fibers and DC cables can be used to replace a large number of RF cable connections between the antenna and other equipment, which greatly simplifies the difficulty of construction and operation. And the introduction of active antenna technology creates conditions for the centralized and cloud processing of baseband. More importantly, the appearance of a large number of controllable antenna ports in the two-dimensional planar array creates the possibility for the system to adjust the beam more flexibly in the three-dimensional space.

For feedback methods based on channel reciprocity, the calibration problem is particularly important. And some channels or elements of the large-scale antenna arrays may fail. Therefore, large-scale antenna systems should have corresponding detection and fault tolerance design mechanisms to ensure the reliability of Massive MIMO transmission. The calibration, monitoring, and fault tolerance design of the antenna array can learn from some mature technologies in the active phased array radar system.

In the high-frequency band system, due to the influence of many unfavorable factors, wireless signal coverage will be more challenging. In this case, a large-scale antenna array can be used to form a beam with high directivity and high gain to overcome many non-ideal factors in signal transmission to ensure coverage distance and transmission quality. Therefore, large-scale antennas have great value for the application and promotion of high-frequency communication technologies.

Compared with low-frequency teaching, the design of large-scale antenna beamforming schemes in high-frequency bands needs to consider some special factors. Due to the complexity, cost, and power consumption, digital-analog hybrid shaping or even pure analog shaping will become the main consideration for system design. In this case, in addition to the digital domain complex baseband channel state information, the system also needs to consider the search and tracking of analog beams and the fast recovery mechanism when blocking occurs. To meet the coverage requirements, the high-frequency destruction system is more suitable for deployment in scenarios where the LOS path is the mainstay. The use of high-gain beams reduces the channel frequency selectivity, so the improvement effect of frequency-selective precoding/beam profile pair performance has been reduced.

Massive MIMO for 6G

Principle of Massive MIMO

The multi-antenna technology has been widely used in almost all mainstream wireless access systems because this technology has great advantages in improving peak rate, system spectrum utilization efficiency and transmission reliability. MIMO technology plays an important role in improving the peak rate and reliability of data transmission, expanding coverage, suppressing interference, increasing system capacity and improving system throughput. The performance gain of the MIMO technology comes from the spatial freedom of the multi-antenna channels. Therefore, expanding the MIMO dimensions has always been an important direction for the standardization and industrialization of the technology. The future mobile communication systems will face greater technical challenges with the rapidly increase in data transmission services and the number of users. The theory of large-scale antenna beamforming came into being under the background of this technological development.

Professor Marzetta from Bell Labs proposed that massive antenna arrays can be used in base stations to form Massive MIMO system to greatly increase the capacity of the system. Massive MIMO refers to the MIMO technology that uses large-scale antenna arrays, as shown in Fig. (**10**). The large-scale arrays are commonly deployed on the top of towers, which could provide scanning beams according to multiple UEs. The coverage of Massive MIMO antenna includes the outdoor scene and the indoor one. It should be mentioned that the millimeter wave AAS could be the promising candidate for the large-scale array in 6G system. And the design idea is similar to spread spectrum communication. The transmitter uses pseudo-random sequences to make the signal tend to be self-contained in the spread spectrum communication. Therefore, the signal can be hidden in noise and interference with a very low SINR and can be detected by the receiver. However,

Massive MIMO uses a large-scale array to whiten the spatial distribution of the signal. As the number of base station antennas increases, the channel coefficient vectors of each user gradually become orthogonal. And Gaussian noise and uncorrelated inter-cell interference are ignorable. Therefore, the number of users that can be accommodated in the system has increased dramatically, and the power allocated to each user can be arbitrarily small. The research results show that the average capacity of each cell can also be as high as 1800Mbit/s with no cooperation between cells and only simple MRC/MRT (Maximum Ratio Combining/Maximum Ratio Transmission) being used for reception/transmission in a 20 MHz bandwidth co-frequency multiplexing TDD system if the base station is equipped with 400 antennas and each cell uses MU-MIMO to serve 42 users. The reason why Massive MIMO obtains huge gains can also be explained from the perspective of beam morphology. As the array scale becomes infinite, the beam formed on the base station side will become very narrow, which will have extremely high directional selectivity and shaping gain. The multi-user interference between multiple UEs will tend to be infinitely small in this case.

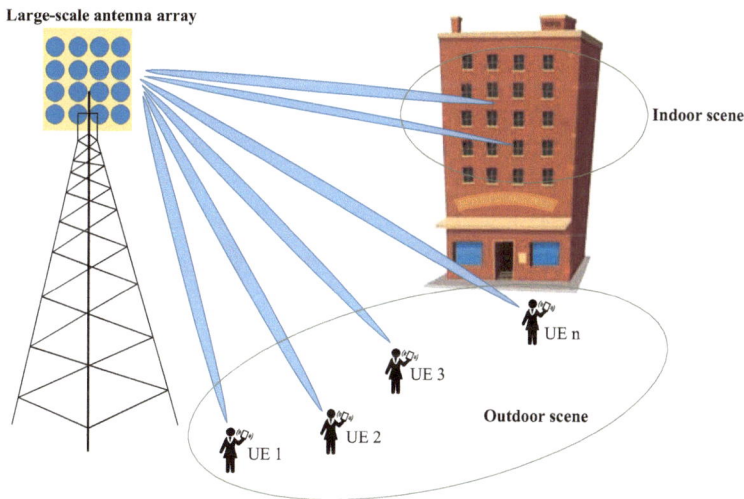

Fig. (10). Massive MIMO with the large-scale antenna array.

Massive MIMO technology immediately became a hot spot in academia and industry after it was proposed. Bell Labs, Lund University in Sweden, Linkoping University in Sweden and Rice University in the United States led the international academic community to conduct extensive explorations on the basic theoretical and technical issues of Massive MIMO channel capacity, transmission, detection and CSI acquisition from 2010 to 2013. The academic community has also actively carried out the principal verification work for the Massive MIMO on the basis of theoretical research. The Lund University published its analysis results based on the measured data of large-scale antenna channels in 2011. The base station of the test system uses a two-dimensional array of 128 antennas, which consists of 16 dual-polarized circular microstrip antennas in 4 rows and the user uses a single antenna. The actual measurement results of the channel show that 98% of the optimal DPC (Dirty Paper Coding) capacity can be achieved even using ZF or MMSE linear precoding when the total number of antennas exceeds 10 times the number of users. These results confirm that the multi-user channel has orthogonality when the number of antennas reaches a certain number. Furthermore, it can ensure that the optimal DPC capacity can still be approached when linear precoding is used. Therefore, the feasibility of Massive MIMO is verified. Rice University, Bell Labs and Yale University jointly built a principal verification platform (Argos) based on 64-antenna arrays in 2012, which can support 15 single-antenna terminals for MU-MIMO. According to the measured data of the received signal, multi-user interference and noise after beamforming, the sum capacity of the system can reach 85 bit/(s. Hz). And it can also reach 6.7 times the spectral efficiency of the SISO system when the total power is 1/64.

The development history of MIMO technology probably follows single-user MIMO, multi-user MIMO, multi-cell multi-user MIMO to massive multi-user MIMO. Many researchers use random matrix theory to theoretically analyze the diversity gain, multiplexing gain and array gain of Massive MIMO for single-user MIMO. People have studied multi-user MIMO in order to make full use of the gains brought by the use of Massive MIMO in base stations considering the limited number of antennas in user terminals, including downlink MIMO-Broadcast Channel (MIMO-BC) capacity and uplink Media Access Control (MIMO-MAC) capacity. The results show that multi-user MIMO can significantly increase the total system capacity, which lays a theoretical foundation for multi-user MIMO transmission. How to expand to multi-cell application scenarios is a problem of MIMO technology in the actual cellular mobile communication system. Therefore, people have studied multi-cell and multi-user MIMO and especially the spectrum efficiency of the system in the presence of multi-cell interference has been studied. Cooperative MIMO technology is proposed to solve the problem of multi-cell interference.

These technologies have been applied in 4G mobile communication systems. However, MU-MIMO and cooperative MU-MIMO have bottlenecks in improving the efficiency of spectrum and power with regard to the typical antenna configuration and cell settings in the current 4G system. The spectrum utilization rate of MU-MIMO in the existing 4G system has not reached the industry's expectations limited by the bottleneck of channel information acquisition and information exchange between base stations. In particular, since the base station is equipped with fewer antennas and the spatial resolution is limited, the system cannot use DPC with an available capacity due to complexity constraints. Therefore, the performance gain of MU-MIMO in the 4G system is still limited.

The spatial resolution of Massive MIMO channels is significantly enhanced and its new features are worth exploring. The base station needs instant channel information when the existing multi-user MIMO technology is directly applied to Massive MIMO scenarios. And there are two problems that are difficult to overcome. The one is that there is a bottleneck in channel information acquisition because the pilot overhead linearly increases with the number of users, moving speed and carrier frequency. The other is that multi-user joint transmission and reception involve matrix inversion, which increases the complexity of the cube and high implementation complexity. Therefore, the Massive MIMO configuration severely restricts the channel information available to the transceiver and the channel capacity analysis needs to consider the channel characteristics and actual constraints.

Fundamental Theory of Massive MIMO

Due to the significant impact of channel environment on the Massive MIMO performance, the capacity of massive MIMO communication system under ideal channel conditions is firstly analyzed based on the theoretical model. The cellular communication system based on Massive MIMO is shown in Fig. (**11**). There are cells in the system, and each cell contains user equipment. antennas are deployed on each base station in every cell. It is assumed that the frequency reuse factor of system is 1, which means that all of the cells work in the same frequency band. OFDM is applied in both uplink and downlink for simplicity. Therefore, the Massive MIMO performance is discussed with single subcarrier.

The channel matrix of the user K in the cell j and the base station in cell is defined as $g_{i,j,k}$

$$g_{l,j,k} = \sqrt{\lambda_{l,j,k}} h_{l,j,k} \tag{4}$$

where $\lambda_{i,j,k}$ is the large scale fading, and $h_{i,j,k}$ is the small scale fading from the user k to the base station in cell I. It can be found that $g_{i,j,k}$ is the M X 1 vector. The small scale fading is assumed as Rayleigh Fading. Thus, the channel matrix from all the k users in cell j to all the base station antennas in cell is:

$$G_{l,j} = \begin{bmatrix} g_{l,j,1} \cdots g_{l,j,k} \end{bmatrix} \tag{5}$$

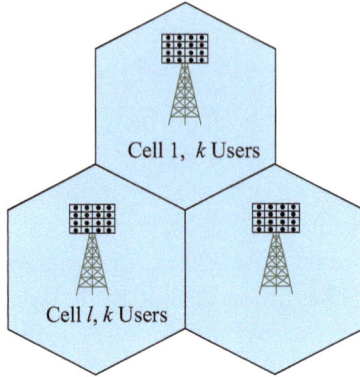

Cell 1, *k* Users

Cell *l*, *k* Users

Fig. (11). Scheme of Massive MIMO for cellular network.

Based on the channel model of Massive MIMO above, the uplink signal received in cell *l* is

$$y_l = G_{l,j}x_l + \sum_{j \neq l}' G_{l,j}x_j + z_l \tag{6}$$

x_1 is the signal transmitted by *k* users in cell *l* , which obeys cyclo-symmetric complex Gaussian distribution z_1 is the Additive White Gaussian Noise Vector. It is shown that Equation **(6)** is the linear model. According to the multi user joint detection with MMSE (Minimum Mean Squared Error), the estimation of transmitted signal is

$$\hat{x}_l = G_{l,l}^H \left(\sum_{j=1}^{L} G_{l,j}G_{l,j}^H + \gamma_{UL}I_M \right)^{-1} y_l \tag{7}$$

The capacity of cell l uplink multiple access is defined as:

$$C = \mathcal{H}\left(x_l \middle| G_{l,1}, \cdots, G_{l,L}\right) - \mathcal{H}\left(x_l \middle| y_l, G_{l,1}, \cdots, G_{l,L}\right) \tag{8}$$

$$\mathcal{H}\left(x_l \middle| G_{l,1}, \cdots, G_{l,L}\right) = \log \det\left(\pi e I_k\right) \tag{9}$$

If $G_{1,1},...,G_{1,L}$ and received signal y_l are known, the uncertainty of can be determined by the error of MMSE detection. Thus, H is rewritten as

$$\mathcal{H}\left(x_l \middle| y_l, G_{l,1}, \cdots, G_{l,L}\right) \le \log\det\left\{\pi e\left[I_k + G_{l,l}^H\left(\sum_{j\neq l}^{L} G_{l,j}G_{l,j}^H + \gamma_{UL}I_M\right)^{-1}G_{l,l}\right]^{-1}\right\} \quad (10)$$

The lower bound of the capacity is consequently

$$C_{LB} = \log\det\left(\sum_{j=1}^{L} G_{l,j}G_{l,j}^H + \gamma_{UL}I_M\right) - \log\det\left(\sum_{j\neq l}^{L} G_{l,j}G_{l,j}^H + \gamma_{UL}I_M\right) \quad (11)$$

According to the large numbers theorem, we can get

$$\lim_{M\to\infty}\frac{1}{M}g_{l,j,k}^H g_{l,j,k'}^H = \begin{cases} \lambda_{1,l,k} & j=1 \quad \text{and} \quad k=k' \\ 0 & \text{otherwise} \end{cases} \quad (12)$$

When the number of base station antennas tends to infinity, the limitation is simplified as

$$\lim_{M\to\infty}\frac{1}{M}G_{l,l}^H G_{l,j} = \begin{cases} \Lambda_{l,l} & j=1 \\ 0 & \text{otherwise} \end{cases} \quad (13)$$

Based on the matrix identity, the capacity of cell l without inter cell interference is

$$C_{LB}^{\inf} = \sum_{k=1}^{K}\log\left(1 + \frac{M}{\gamma_{UL}}\lambda_{l,l,k}\right) \quad (14)$$

According to the analysis above, if the receiver knows the ideal channel information, multi-user interference and multi cell interference disappear. The system capacity increases logarithmically with the number of antennas, and tends to infinity. The same conclusion could be obtained with the MRC (Maximum Ratio Combination).

The factors that affect the performance of large-scale antenna system in practical application are studied and analyzed. Firstly, it is necessary to evaluate the performance of large-scale antenna systems in the complex wireless channels. The direct path in Rice Fading channel and the drastic change of channel state caused by mobile terminal movement will have a negative impact on the performance of large-scale antenna system. Secondly, the influence of non ideal factors of hardware equipment on system performance should be considered with the engineering application of large-scale antenna system. It is necessary to analyze the relationship between non ideal reciprocity and capacity in TDD (Time Division Duplex) system. Finally, the system level capacity of large-scale antenna cellular system and the relationship between system parameters and unit area capacity of cellular network will be evaluated.

This section analyzes the system capacity of massive MIMO in Rice Fading channel. In the actual communication system, the array element spacing is small attributed to the limited volume of base station equipment. Therefore, there is a high correlation between the base station side antennas. On the other hand, considering that there may exist LOS (Line of Sight) path between users and base stations, Rice Fading channel model is usually used for channel modeling. The uplink is taken as an example to analyze the spectrum effectiveness of the channel estimation scheme.

Considering the system model shown in Fig. (**11**), the cell is set as observation cell. Equation (**6**) could be rewritten as:

$$y = G_1 x_1 + \sum_{l=2} G_l x_l + z \tag{15}$$

Since the distance to the target base station is relatively close, the fast fading modeling between the user terminal and the service base station consists of two parts for the user terminal of the target cell: the deterministic component generated by LOS propagation and the random component generated by Rayleigh Distribution. Meanwhile, the direct path component can be ignored due to the scattering, diffraction and diffraction of many obstacles for the user who interferes with the cell. The fast fading part of the channel can be modeled as:

$$H_l = \begin{cases} \bar{H}_1 \left[\Omega(\Omega + I_k)^{-1} \right]^{\frac{1}{2}} + \hat{H}_1 \left[(\Omega + I_k)^{-1} \right]^{\frac{1}{2}} & l = 1 \\ \hat{H}_l & l \neq 1 \end{cases} \tag{16}$$

where $\bar{\boldsymbol{H}}_1$ is the LOS component, and $\hat{\boldsymbol{H}}_l$ is the random component generated by Rayleigh Distribution. $\boldsymbol{\Omega}$ is a diagonal matrix, of which the diagonal elements $\boldsymbol{\Omega}_{k,k}$ are the Rice factor of target cell user k. Considering the base station side antenna correlation and assuming that all users have the same correlation to the target base station, the total channel matrix is:

$$\boldsymbol{G}_l = \begin{cases} \bar{\boldsymbol{G}}_1 \left[\boldsymbol{\Omega}(\boldsymbol{\Omega}+\boldsymbol{I}_k)^{-1} \right]^{\frac{1}{2}} + \hat{\boldsymbol{G}}_1 \left[(\boldsymbol{\Omega}+\boldsymbol{I}_k)^{-1} \right]^{\frac{1}{2}} & l = 1 \\ \hat{\boldsymbol{G}}_l & l \neq 1 \end{cases} \tag{17}$$

$$\begin{aligned} \bar{\boldsymbol{G}}_1 &= \left[\bar{\boldsymbol{g}}_{1,1}, \bar{\boldsymbol{g}}_{1,2} \cdots \bar{\boldsymbol{g}}_{1,k} \right] = \bar{\boldsymbol{H}}_1 \boldsymbol{\Lambda}_1^{\frac{1}{2}} \\ \hat{\boldsymbol{G}}_l &= \left[\hat{\boldsymbol{g}}_{1,1}, \hat{\boldsymbol{g}}_{1,2} \cdots \hat{\boldsymbol{g}}_{1,k} \right] = \boldsymbol{R}^{\frac{1}{2}} \hat{\boldsymbol{H}}_l \boldsymbol{\Lambda}_l^{\frac{1}{2}} \end{aligned} \tag{18}$$

If the exact Los determination component and rice factor matrix $\boldsymbol{\Omega}$ are known by the user terminal and the target base station, only the channel estimation of Rayleigh fading is required to be considered. The channel matrix model with channel estimation is defined with MMSE.

$$\hat{\boldsymbol{g}}_{1,k} = \begin{cases} \dfrac{\lambda_{1,k}\boldsymbol{R}}{\vartheta_k+1} \boldsymbol{Q}_k^{\frac{1}{2}} \hat{\boldsymbol{h}}_k + \dfrac{\sqrt{\vartheta_k}}{\sqrt{\vartheta_k+1}} \bar{\boldsymbol{g}}_{1,k} & l = 1 \\ \lambda_{1,k}\boldsymbol{R}\boldsymbol{Q}_k^{\frac{1}{2}} \hat{\boldsymbol{h}}_k & l \neq 1 \end{cases} \tag{19}$$

$$\boldsymbol{Q}_k = \left(\frac{\lambda_{1,k}\boldsymbol{R}}{\vartheta_k+1} + \sum_{l=2}^{L} \lambda_{1,k}\boldsymbol{R} + \gamma_P \boldsymbol{I}_M \right)^{-1} \tag{20}$$

According to the orthogonality of MMSE estimation, the covariance matrix of estimation error is:

$$\operatorname{cov}\left(\tilde{\boldsymbol{g}}_{l,k}, \tilde{\boldsymbol{g}}_{l,k} \right) = \begin{cases} \dfrac{\lambda_{1,k}\boldsymbol{R}}{\vartheta_k+1} - \left(\dfrac{\lambda_{1,k}}{\vartheta_k+1} \right)^2 \boldsymbol{R}\boldsymbol{Q}_k\boldsymbol{R} & l = 1 \\ \lambda_{l,k}\boldsymbol{R} - \lambda_{l,k}^2 \boldsymbol{R}\boldsymbol{Q}_k\boldsymbol{R} & l \neq 1 \end{cases} \tag{21}$$

The channel estimation results show that the observed signal is:

$$y = \sum_{i=1}^{K} \hat{g}_{1,i} x_{1,i} + \sum_{i=1}^{K} \tilde{g}_{1,i} x_{1,i} + \sum_{l=2}^{L}\sum_{i=1}^{K} \hat{g}_{l,i} x_{l,i} + \sum_{l=2}^{L}\sum_{i=1}^{K} \tilde{g}_{l,i} x_{l,i} + z \tag{22}$$

Based on the Linear detection scheme and MRC filter, the SINR (Signal to interference noise ratio) of the user k can be expressed as:

$$\Gamma_k = \frac{\left| \hat{g}_{1,k}^{H} \hat{g}_{1,k} \right|^2}{\hat{g}_{1,k}^{H} \left(\sum_{l=1}^{L}\sum_{i \neq k}^{K} \hat{g}_{l,i} \hat{g}_{l,i}^{H} + \sum_{l=1}^{L}\sum_{i=1}^{K} \tilde{g}_{l,i} \tilde{g}_{l,i}^{H} + \gamma_{UL} I_M + \sum_{l=2}^{L} \hat{g}_{l,k} \hat{g}_{l,k}^{H} \right) \hat{g}_{1,k}} \tag{23}$$

When the number of base station antennas tends to infinity, the received SINR of the user k in the target cell will approach to:

$$\Gamma_k^{\infty} = \frac{1}{\sum_{l=2}^{L} \lambda_{l,k}^2} \left(\frac{\lambda_{1,k}}{\vartheta_k + 1} + \frac{M\vartheta_k}{\sum_{n=1}^{M} \delta_{k,n}^2} \right)^2 \tag{24}$$

It can be found that when the number of base station antennas tends to infinity, the irrelevant interference and noise are eliminated. The interference components are just from other cell users using the same pilot. The signal power of the target user includes Los component and Rayleigh fading component. Based on the analysis of influence of LOS component on received SINR, the transmitted and received signals are modeled as:

$$x = \frac{\lambda_{1,k}}{\vartheta_k + 1} + y$$

$$y = \frac{M\vartheta_k}{\sum_{n=1}^{M} \dfrac{d_n^2}{\dfrac{\lambda_{1,k} d_n}{\vartheta_k + 1} + d_n \sum_{l=2}^{L} \lambda_{1,k} + \gamma_P}} \tag{25}$$

Equation **(25)** shows that increases with. It means that the direct path component of the target user channel improves the user SINR. It is found from Equation **(24)** that the size of LOS component in the target user signal power is related to the antenna correlation at the base station side. Furthermore, the size of Rayleigh fading component is independent from the antenna correlation at the base station side. When the Rice factor of all users is the same and tends to infinity, the user SINR is given by Equation **(26)**:

$$\Gamma_k \to \frac{\lambda_{1,k} M^2}{\displaystyle\sum_{l=2}^{L}\sum_{i=1}^{K} \frac{\lambda_{l,i}}{\lambda_{1,k}} \bar{\boldsymbol{g}}_{1,k}^{H} \boldsymbol{R} \boldsymbol{g}_{1,k}^{H} + \sum_{i\neq k}^{K} \lambda_{1,i} \left|\rho_{k,i}\right|^2 + \gamma_{UL} M} \tag{26}$$

Equation **(26)** shows that the signal power of the target user increases with the square of the number of antennas on the base station side, and the interference and noise power of other cell users increases linearly with the number of base station side antennas. When the number of antennas tends to infinity, the noise interference, other cell user interference and the interference of other users in the target cell are gradually eliminated. If the LOS components of all the users are exactly estimated, the target user rate increases linearly with the number of base station antennas.

How to improve the high data transmission rate under high mobility has become one of the difficult problems in mobile communication with the increasing demand of data transmission rate in high-speed railway and highway scenarios. In order to solve the bottleneck problem of data transmission rate in high-speed mobile scene, large-scale antenna is the main solution.

In the high-speed scenario, due to the relative motion between the mobile terminal and the scattering environment in the wireless channel, the signals arriving at the receiving end at different times experience different degrees of fading, that is, the time selective fading of the channel, which aggravates the time-varying of the channel. Thus, there are errors in the channel estimation obtained from the base station side, which will affect the performance of large-scale antenna system.

This section analyzes Massive MIMO in time-varying channel environment. According to the equivalent channel model, the closed expressions of the lower bound of uplink and rate are derived when MMSE linear receiver is used in multi-cell multi-user MIMO system. The asymptotic expression of the lower bound is given when the number of base station antennas tends to infinity. The asymptotic expression is used to analyze the channel delay correlation The limitation expression of uplink sum rate is given, when the ratio of base station antenna number to user number tends to infinity.

Base on the communication scene shown in Fig. (**11**), the uplink is modeled with time varying characteristics of channel

$$y(t) = G_1(t) x_l(t) + \sum_{l \neq 1}^{L} G_l(t) x_l(t) + z(t) \tag{27}$$

According to the equivalent signal model, the combined rate can be expressed as

$$
\begin{aligned}
C^{\tau} = \log_2 \det \left[\sum_{i=1}^{L} \hat{G}_i^{\tau} \left(\hat{G}_i^{\tau} \right)^H \left(\Sigma_i^{\tau} \right)^{-1} + I_M \right] - \\
\log_2 \det \left[\sum_{i \neq 1}^{L} \hat{G}_i^{\tau} \left(\hat{G}_i^{\tau} \right)^H \left(\Sigma_i^{\tau} \right)^{-1} + I_M \right]
\end{aligned}
\tag{28}
$$

If the number of antennas tends to infinity, it is found

$$C^{\tau} - C_{\text{inf}}^{\tau} \to 0$$

$$C_{\text{inf}}^{\tau} \to \sum_{k=1}^{K} \log_2 \left[1 + \frac{\left| \lambda_{1,k} \rho_{1,k}^{\tau} \right|^2}{\sum_{i \neq 1}^{L} \left| \lambda_{i,k} \rho_{i,k}^{\tau} \right|^2} \right] \tag{29}$$

When the number of antennas tends to infinity in the high-speed mobile scene, the influence of Gaussian noise, channel delay related estimation error and fast fading are completely eliminated, and the inter user interference in the target cell is also completely removed. However, due to pilot pollution, the interference of users using the same pilot in other interference cells still exists. The signal strength and interference signal strength of the target cell are proportional to the square of the channel delay coefficient.

If the mobile speed of the user in the scheduling target cell is lower than that of the user in the interfering cell, the uplink sum rate of the target cell will increase instead of decreasing compared with that of all users. If the mobile speed of the target cell is higher than that of the interfering cell, the sum rate of the target cell must be reduced. If the users of all cells are scheduled to move at the same speed, the system and rate of all cells will not be reduced due to the user movement. These conclusions have important value for mobile communication system architecture design. The reason for these phenomena is that the channel delay not only affects the useful signal power, but also affects the adjacent cell user interference caused by pilot pollution.

In large-scale antenna system, with the increase of the number of base station antennas and space separation users, the channel information acquisition becomes the bottleneck of system implementation. When TDD mode is adopted, the base station can use the uplink channel estimation information to design downlink precoding in coherent time, thus reducing the overhead of downlink pilot and CSI feedback. However, the whole channel for the actual system includes not only the wireless part in the air, but also the RF circuit of the transceiver. Although the air channel satisfies the reciprocity between uplink and downlink, the overall channel of uplink and downlink cannot guarantee the reciprocity accuracy without precise circuit calibration. It means that the inconsistency of the transceiver RF circuit should be carefully considered. In this section, the influence of non-ideal reciprocity on the performance of massive MIMO system is studied through theoretical analysis.

Considering the RF circuit gain at both ends of a complete communication channel, the channel matrix of uplink and downlink is

$$
\begin{aligned}
G_{UL} &= C_{BS,r} H^T C_{UE,t} \\
G_{DL} &= C_{UE,r} H C_{BS,t}
\end{aligned}
\tag{30}
$$

where H is the downlink channel matrix, of which element is the channel gain of antenna m and user k. The elements of H are Cyclic symmetric complex Gaussian random variables with independent and identically distributed zero mean unit variance. $C_{BS,t}$ and $C_{BS,r}$ are RF gain matrix corresponding to the transmitting and receiving of the base station respectively. $C_{UE,t}$ and $C_{UE,r}$ are RF gain matrix corresponding to the transmitting and receiving of the user equipment respectively. According to the channel model, due to the RF gain mismatch between the base station and the user, the reciprocity between the uplink and downlink channel matrices is no longer tenable, that is $G_{DL} \neq G_{UL}$.

It is assumed that the base station side can obtain the ideal uplink channel state information for simplicity. If ZF precoding is used for signal transmission, the downlink signal received by the terminal is

$$y = \beta_{mis} \boldsymbol{G}_{DL} \boldsymbol{G}_{UL}^* \left(\boldsymbol{G}_{UL}^T \boldsymbol{G}_{UL}^* \right)^{-1} \boldsymbol{x} + \boldsymbol{n} \qquad (31)$$

where β_{mis} is coefficient which normalizes the transmission power and the RF gain in the downlink channel matrix. According to the matrix representation of uplink and downlink channels, the received signal is:

$$y = \beta_{mis} \boldsymbol{C}_{UE,r} \boldsymbol{W} \boldsymbol{C}_{UE,t}^{-1} \boldsymbol{x} + \boldsymbol{n}$$
$$\boldsymbol{W} = \left(\boldsymbol{H} \boldsymbol{C}_{BS,t} \boldsymbol{C}_{BS,r}^* \boldsymbol{H}^H \right) \left(\boldsymbol{H} \boldsymbol{C}_{BS,r} \boldsymbol{C}_{BS,r}^* \boldsymbol{H}^H \right)^{-1} \qquad (32)$$

Equation **(32)** shows that the matrix $\boldsymbol{W}$ is not an identity matrix, due to the RF gain mismatch at base station side. Therefore, the mismatch of RF circuit gain between transmitter and receiver will destroy the reciprocity of communication channel, which will lead to interference between users. On the other hand, through similar methods, it can be found that terminal RF gain mismatch has little impact on system performance, and phase mismatch does not affect system throughput.

System level spectrum efficiency is an important index for industry to evaluate cellular network mobile communication system, which is usually obtained by very complex and time-consuming system level simulation. In recent years, in order to get the relationship between system level spectrum efficiency and system parameters theoretically, a lot of research has been done.

There are generally two methods to study the system level spectrum efficiency. The first method is based on the random geometric model. According to the SINR of the receiver, the spectrum efficiency is obtained by Shannon formula, and then the spectrum efficiency of the system is obtained by assuming that the base station deployment follows Poisson distribution. However, the closed expression of system level spectral efficiency obtained by Poisson distribution is very complex, and it is difficult to give the relationship between system spectral efficiency and system parameters directly. In addition, the analysis of system level spectrum efficiency is rare due to the complexity of the expression of SINR, considering the non-ideal channel information.

The second system level spectrum efficiency analysis method assumes that the base station location is fixed and known, and users are evenly distributed in the cell. The system level spectrum efficiency is obtained based on the expectation of the user location. The channel model is simplified with the assumption:

$$\lambda_{l,k} = cC_{ASE} = \frac{(T-K)K}{\pi R^2 T} \log_2\left[1 + \frac{\phi_2(d_1)}{\Delta_1 + \Delta_2 K + \Delta_3/K}\right]$$

$$\Delta_1 = \frac{\phi_1(d)\gamma}{MP_D}\left(\frac{1}{\beta}+1\right) + \phi_2(d_1)$$

$$\Delta_2 = \frac{1}{M}\left[\phi_1^2(d) - \phi_2(d)\right] \tag{33}$$

$$\Delta_3 = \frac{\gamma^2}{\beta MP_D^2}$$

where is the distance from user in cell to base station in cell 1, and is the path loss index. Thus, the system capacity is

$$C = \sum_{k=1}^{K}\log_2\left[1 + \frac{d_{1,k}^{-2\alpha}}{\sum_{l=2}^{L}d_{l,k}^{-2\alpha} + \frac{\varepsilon + \gamma/P_D}{M}\left(\sum_{l=1}^{L}d_{l,k}^{-\alpha} + \frac{\gamma}{KP}\right)}\right] \tag{34}$$

is the transmitted power per data symbol, and the total pilot power of each user is. The Reference Signal to Noise Ratio is defined as to evaluate system level performance. Since there are many users served simultaneously at the same frequency in Massive MIMO system, the system capacity is approximately written as:

$$C = K\varepsilon\left\{\log_2\left[1 + \frac{d_1^{-2\alpha}}{\sum_{l=2}^{L}d_l^{-2\alpha} + \frac{\varepsilon + \gamma/P_D}{M}\left(\sum_{l=1}^{L}d_l^{-\alpha} + \frac{\gamma}{KP}\right)}\right]\right\} \tag{35}$$

The spectral efficiency per unit area is defined as:

$$C_{ASE} = \frac{T-\tau}{\pi R^2 T} C \qquad (36)$$

where is the pilot loss, is the number of symbols transmitted in coherent time. Therefore, the spectral efficiency per unit area is:

$$
\begin{aligned}
C_{ASE} &= \frac{(T-K)K}{\pi R^2 T} \log_2 \left[1 + \frac{\phi_2(d_1)}{\Delta_1 + \Delta_2 K + \Delta_3 / K} \right] \\
\Delta_1 &= \frac{\phi_1(d)\gamma}{MP_D}\left(\frac{1}{\beta}+1\right) + \phi_2(d_1) \\
\Delta_2 &= \frac{1}{M}\left[\phi_1^2(d) - \phi_2(d)\right] \\
\Delta_3 &= \frac{\gamma^2}{\beta M P_D^2}
\end{aligned}
\qquad (37)
$$

It can be found that the spectrum efficiency of large-scale antenna does not increase with the increase of K, considering the pilot overhead. the pilot overhead increases linearly with the increase of K. Therefore, there will be an optimal number of users supported by the system with the given coherent time. According to Equation **(37)**, the optimal number of users can be obtained by searching. In particular, in the low SNR region, Equation **(37)** can be further simplified as:

$$C_{ASE} = \frac{(T-K)K}{\pi R^2 T \ln 2}\left[1 + \frac{\phi_2(d_1)}{\Delta_1 + \Delta_2 K + \Delta_3 / K} \right] \qquad (38)$$

The optimal number of users can be obtained by solving the cubic Equation (39) with $\frac{\partial C_{ASE}}{\partial K} = 0$

$$\Delta_2 K^3 + 2\Delta_1 K^2 + \left(3\Delta_3 - T\Delta_1\right)K - 2T\Delta_3 = 0 \qquad (39)$$

In addition, the pilot data power ratio can also be analyzed with Equation (37), when the spectrum efficiency reaches the maximum. The energy required by the user to transmit all pilot symbols is during the symbol time, and energy consumption of data transmission is. Considering the actual communication system, the transmission power of each user will be limited by a maximum value

$$\frac{1}{T}\left[KP+(T-K)P_D\right] \le P_{\max} \tag{40}$$

According to the Equation **(37)**, the analysis of system level parameters can be transformed into optimization mathematical problems with given number of users, number of base station antennas and coherent time.

$$\min_{P_D,P}\left\{\frac{\phi_1(d)}{P}+\frac{\phi_1(d)}{P_D}+\frac{\gamma}{KPP_D}\right\}$$

$$\text{s.t.}\quad \frac{1}{T}\left[KP+(T-K)P_D\right] \le P_{\max} \tag{41}$$

$$P>0$$

$$P_D>0$$

The above objective function in Equation **(41)** is convex function. Therefore, KKT (Karush Kuhn Tucker) condition is a necessary and sufficient condition for the optimization problem. According to Lagrange multiplier method, the optimal pilot power and data power can be obtained, and then the optimal pilot data power ratio can be obtained.

RIS-Aided Wireless Communications

In view of the demand of 6G communication system for faster and more reliable data transmission technology, RIS has been widely and deeply studied by academia and industry [7]. RIS is generally a planar surface containing many passive reflecting units, each of which could control the amplitude and/or phase of the scattering field independently. By densely deploying RIS in communication system and smartly regulating the scattered field, the propagation channels between transmitters and receivers can be flexibly reconfigured to achieve desired realizations, which thus provides a novel approach to tackle the channel fading impairment and interference issue. Since the change of scattered field is currently considered as a phase shift, the RIS consumes no transmit power.

Reconfigurable Intelligent Surface (RIS)

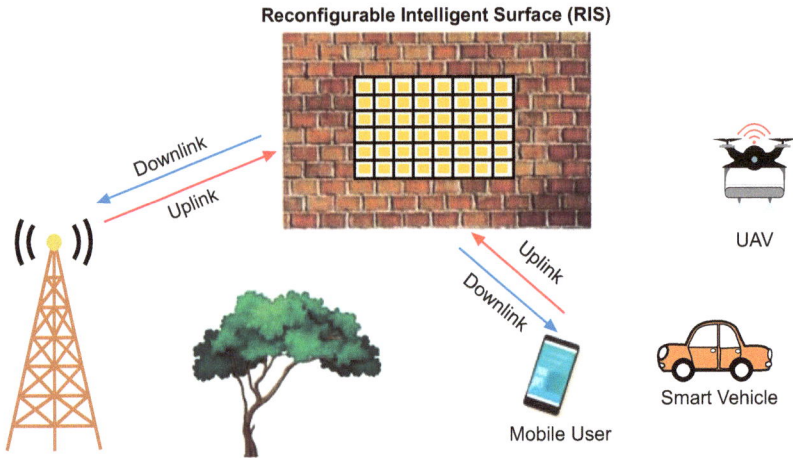

Fig. (12). Scheme of RIS-aided communication between a base station and a mobile user/UAV/smart vehicle.

As shown in Fig. (**12**), the RIS equipment is deployed between the base station and user equipment. RIS deployment scenarios are generally planar structure carrier, such as walls, building facades and ceilings. The LoS propagation of wireless signal is interfered by the trees, vegetation, buildings and other obstacles. The user equipment contains mobile phone, unmanned aerial vehicle (UAV), a smart vehicle, or any other terminal. The performance of communication system is effectively enhanced by the modulation of transmission signal by RIS.

RIS is similar to massive MIMO technology, such as using large-scale antenna array to improve spectrum efficiency and energy efficiency. However, RIS improve the propagation environment for the communication system. It has been proved that the capacity which can be enhanced per square meters surface area has a linear relationship with the average transmitted power, instead of being logarithmic in the case of the Massive-MIMO. Moreover, compared with the recent technologies, such as backscatter communication, millimeter wave communication and network densification, RIS could control the mobile environment and basically consume much less power. As a newly developing communication technology, RIS has also been given other names in the literature, such as large intelligent surface (LIS), large intelligent metasurface (LIM), intelligent reflecting surface (IRS), software-defined surface (SDS) and passive intelligent surface (PIS). For consistency, in the rest of the paper, we will use the name RIS instead of other terms listed above.

It has been found that RIS possesses various practical advantages for implementation compared with Massive MIMO. The reflecting units of RIS are passive, which only reflect the incident electromagnetic wave without requiring any radio frequency (RF) chains. Furthermore, combination of RIS and full-duplex (FD) could avoid any antenna noise amplification and self-interference, which provides attractive advantages over traditional active communication system. Moreover, due to the compact size and conformal structure, RIS can be easily deployed on/removed from targets for deployment/replacement. Finally, IRS operates as an auxiliary device in mobile networks and can be integrated into them transparently, and providing great flexibility and compatibility with existing mobile systems.

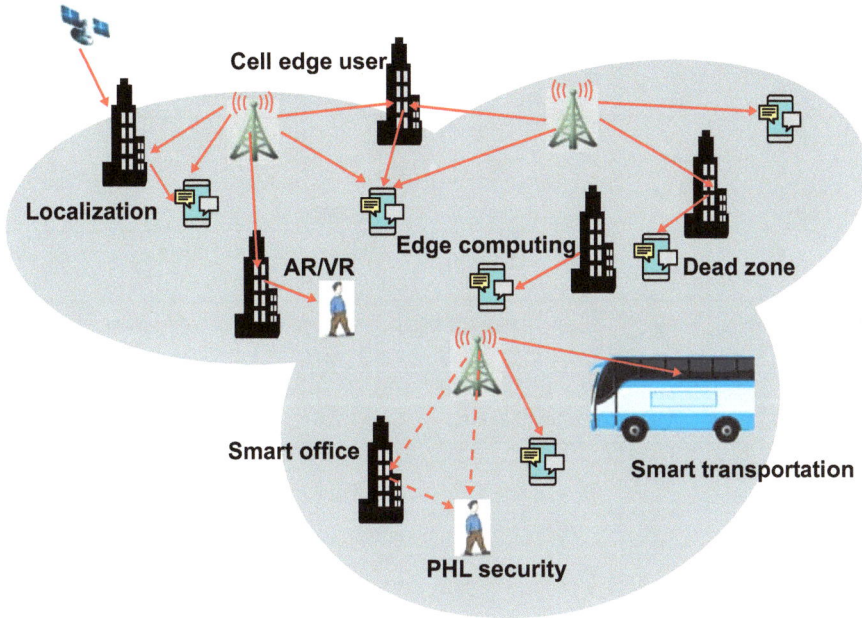

Fig. (13). Illusion of RIS applied in the 6G communication system.

Considering the above promising advantages, RIS could be massively deployed in 6G wireless networks to significantly improve the spectral and energy efficiency cost-effectively. The future wireless network aided by RIS is shown in Fig. (**13**). RIS is deployed to provide a LoS link between the users located in a service dead zone, and their serving base station which bypasses the obstacle between them. This application is useful to extend the cell coverage in high frequency band communications (such as mmWave and THz communication), which may be influenced by blockage. Besides, deploying RIS at the cell edge not only improves the desired signal power at cell-edge users but also facilitates the suppression of neighboring co-channel interference. Moreover, RIS can be utilized to compensate the power loss over long distance with reflect beamforming for the efficiency of simultaneous wireless information. RIS could also be deployed on the building facade, lamppost, advertising board, and even the surface of high-speed moving vehicles in outdoor environment to support URLLC for remote control and smart transportation by effectively compensating the Doppler effects. RIS is a disruptive technology for future intelligent communication environment, which can potentially benefit a wide range of vertical industries in 6G, such as transportation, manufacturing, smart city, *etc.* Thus, RIS has been considered as one promising technology for the 6G ecosystem. In addition, several pilot projects have been launched to advance the research in wireless communication new field, with more information given in Table **1**.

Table 1. List of main industry progress to RIS.

Company	Year	Main Activity and Achievements
Lumotive and TowerJazz	2019	Demonstrate the first true solid-state beam steering using liquid crystal metasurface
Pivotal Commware	2019	Demonstrate holographic beamforming technology using software-defined antennas
NTT DOCOMO and AGC Inc.	2020	Demonstrate the first prototype transparent dynamic metasurface for 5G
Research Project	**Start Year**	**Main Objective**
ARIADE	2019	Design metasurface integrated with new radio and artificial intelligence (AI) techniques
PathFinder	2021	Establish the theoretical and algorithmic foundations for intelligent metasurface enabled wireless 2.0 networks

The recent research work on RIS are classified as followed. The capacity rate analyses, power/spectral optimizations, channel estimation, deep learning-based design, and reliability analysis are discussed in detailed.

A. Capacity Rate Enhancement

Considering the entire RIS surface as a reflecting antenna array, the received signal can be well represented by a sinc-function-like intersymbol interference channel. The capacity per square meters surface-area converges to, when the wavelength tends to zero, where is the transmit power per volume-unit, and denotes the additive white Gaussian noise's spatial power spectral density. Although the capacity is limited by the aperture of RIS surface, splitting the reflecting surface into the array composing many units can alleviate the capacity degradation. It is found that the asymptotic capacity result is in accordance with the exact mutual information as the numbers of devices increase. Considering the discrete phase shifts, the weighted sum of downlink rates could be maximized by optimizing the active beamforming, in which each weight represents the priority of the user.

B. Power Optimizations

When the base station employs a well-designed zero-forcing precoding matrix to achieve perfect interference suppression, the energy efficiency of the downlinks could be maximized by optimizing the RIS phase matrix and power allocation at the base station side. Moreover, the downlink power could also be minimized for an RIS-aided multiple access network. The spectral efficiency is apparently promoted by optimizing the beamforming of RIS surface.

C. Channel Estimation

Based on the minimum mean squared error (MMSE), the total channel estimation time is divided into a sequence of sub-phases. The downlinks are estimated at the base station with control signals, and the channel information is reported by the base station to the RIS controller. Firstly, all RIS units are turned OFF and the base station estimates the direct channels for all users. Then each RIS unit turns to be ON while all other RIS units are OFF to allow estimations by the base station. Finally, estimation results of all sub-phases are taken together to obtain the channel estimation.

D. Deep Learning-Based Design

Compressive deep learning could be utilized for configuring RIS-aided wireless communications. Regarding the wireless propagation as a deep neural network, the wireless network learns the propagation basics of RIS and configures them to the optimal setting. The wireless channel qualities are learned with a deep neural network. Furthermore, deep learning is also used to guide the RIS to learn the optimal interaction with the incidence signals.

E. Secure Communications

RIS could also be used to secure the physical layer of wireless communications. In the future 6G system, the simple wiretap channel model could be extended to the broadcast wiretap channel, compound wiretap channel, Gaussian wiretap channel, and MIMO wiretap channel. In order to secure the communication system, RIS could be applied to increase the data rate at a legitimate user and decrease the one at an eavesdropper. Considering the case of multiple legitimate users and eavesdroppers, the RIS-aided downlink broadcast system consists of a multi-antenna base station, multiple legitimate users with each having single antenna, and multiple eavesdroppers. The minimum secrecy data rate among all legitimate users could be maximized by finding the optimal transmit beamforming and RIS phase shifts with alternating optimization.

F. Terminal-positioning and Other Novel Applications

Based on the beamforming algorithm, terminal-positioning technology for wireless communication could be improved with RIS. Three Cartesian dimensions of a terminal is obtained by the Cram'er–Rao lower bounds (CRLBs). Except for the case of a terminal locating perpendicular to the RIS center, the CRLB degrades quadratically with respect to the RIS surface area. Moreover, RIS has been applied to assist over-the-air computation (AirComp), where the base station is recognized as a computation node. The key procedure is the optimization problem of finding the RIS phase shifts and the decoding vectors to minimize the distortion after signal decoding. An alternating difference-of-convex algorithm has been proposed with majorization-minimization technique for the non-convexity of the problem.

MASSIVE MIMO ANTENNA FOR MOBILE COMMUNICATIONS

For 6G base station equipment with massive MIMO antennas, the overall architecture can adopt the distributed architecture commonly used in traditional commercial macro base station, which is composed of evolved base-band unit (eBBU) and Active Antenna Unit (AAU), which is a base station combination that can be flexibly distributed and installed. As shown in Fig. (**14**), AAU is connected with enhanced base-band unit equipment through Common Public Radio Interface/evolved common public radio interface, and eBBU is connected with core network through NG interface.

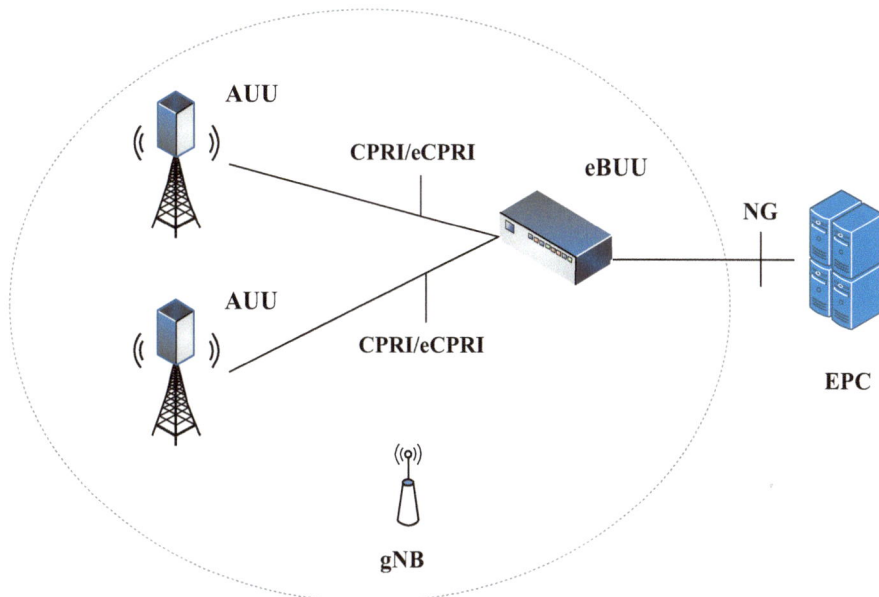

Fig. (14). Active Antenna Unit applied in the wireless communication system.

Compared with the traditional MIMO technology, Massive MIMO technology adopts a large-scale antenna array at the base station (the number of array elements reaches several hundred), thus separating multipath channels more effectively, further concentrating radiated energy, and improving spectrum reuse rate and signal-to-noise ratio. To detect and separate multipath channels, each array element in the MIMO antenna array has an independent radio frequency link, which includes antenna radiating elements, radio frequency amplifiers, up-down converters, digital/analog converters, and so on. Compared with the traditional MIMO antenna array, the complexity, cost, and power consumption of the system will rise sharply. The following subsections show the designs and synthesis of Massive MIMO Antenna element and array, Feed Network and RF Front-end Design, and the theory of Antenna selection.

Massive MIMO Antenna Array Design and Synthesis

Array design is a key link in the realization of massive MIMO antenna equipment. Massive MIMO antenna elements are arranged according to a certain physical structure to form an antenna array to complete the desired shaping function. It is necessary to analyze the constraints and design methods of the array structure from the aspects of feasibility and performance. This section will discuss the structural design of large-scale antenna arrays. From several key factors such as decoupling between antennas, beamforming, feeder network, and RF front-end design, *etc.*, an in-depth analysis of the related issues of the array design, and a commonly used structural design scheme.

Massive MIMO Antenna Decoupling Technology

For an antenna array composed of multiple linear elements, the multiple antennas in the space will be affected by the current on other antennas and the enveloping field. The antenna elements placed close to each other have a turning effect. Since the mutual coupling effect between the antenna elements affects the shaping effect of the antenna beam, a certain distance must be maintained between the large-scale antenna array elements. The miniaturization of modern mobile devices has become a popular trend, but MIMO technology requires that the multiple antennas of the transmitting and receiving end (receiving end) are relatively independent. Since multiple antennas are affected by mutual coupling, it is extremely difficult to make the antennas independent of each other [8]. The existence of mutual coupling effects brings difficulties to the miniaturization of large-scale antenna arrays. Therefore, it is required to reduce the size of the MIMO system under the premise of reducing the coupling effect between antennas in the process of large-scale antenna design. In engineering, isolation or coupling coefficient is usually used to express the coupling between antennas. To reduce the influence of mutual coupling between antennas on antenna performance, common methods for decoupling between multiple antennas include defective ground, parasitic resonant units, neutralization lines, decoupling networks, and new electromagnetic material structures.

Defectively improve the isolation method: Defected Ground Structure (DGS) to form a parallel LC resonant circuit to achieve band rejection characteristics. By changing the current distribution on the metal floor, the surface wave transmission of the antenna is suppressed, and the mutual coupling between the antennas is reduced. The parasitic radiating element is a method for improving the isolation. The parasitic resonant element is loaded to absorb the near-field radiation wave of each antenna of the array, thereby improving the isolation between the antennas.

The neutralization line technology and decoupling network technology are based on the idea of phase compensation to achieve the purpose of reducing coupling. When the two antennas are placed in a limited space, a coupling compensation of the opposite phase is introduced to the coupling between the antennas, thus the isolation between the antennas is realized. The neutralization line technology connects two antennas through the neutralization line. Through reasonable design and optimization, the coupling current on the neutralization line produces the opposite phase with the original current to achieve neutralization. The access point and length of the neutralization line in the antenna play a key role in phase compensation. The decoupling network also cancels the coupling phase of the two antennas based on the idea of phase compensation, to achieve the effect of isolation.

The Electromagnetic Band Gap structure (EBG) can be used to improve the insulation. The classic EBG is a mushroom-shaped structure, which is periodically arranged by a series of metal sheets printed on one side of the dielectric substrate, and these metal sheets are connected to the metal floor on the other side of the dielectric substrate through short-circuiting holes. The electromagnetic band-gap structure can suppress the surface wave. The EBG structure is loaded between the two antennas to suppress the surface wave to restrain the coupling between the antennas to improve the separation degree of the antenna spacing.

Large-Scale Antenna Beamforming Technology

Beamforming technology comes from electromagnetic wave theory, which originated from military radar application at first, and is now gradually adopted by communication technology [9]. The 6G mobile communication system adopts beamforming technology to realize high-speed data transmission, and its system model is shown in Fig. (**15**).

Beamforming is the basis of large-scale antenna work. The performance of beamforming is closely related to the comprehensive design of the array pattern. Pattern synthesis is one of the key issues in antenna design, whether for the traditional mobile communication base antenna or the new generation of the large-scale antenna. Array pattern synthesis is to synthesize the number of array elements, array spacing, and optimize a set of excitation weights according to the spatial radiation characteristics (pattern shape, main lobe width, sidelobe level, directivity coefficient) needed for beamforming. The weight should include the amplitude and phase of each element in the array. Therefore, in the case of a given number of array elements, there are three methods to adjust the radiation pattern of the array antenna: adjusting the excitation amplitude of the array elements, adjusting the excitation phase of the array elements, and adjusting the spatial distribution of the array elements. These three can be used alone or at the same time. According to the given conditions and synthesis methods, the pattern synthesis problems of array antennas can be roughly divided into the following four categories.

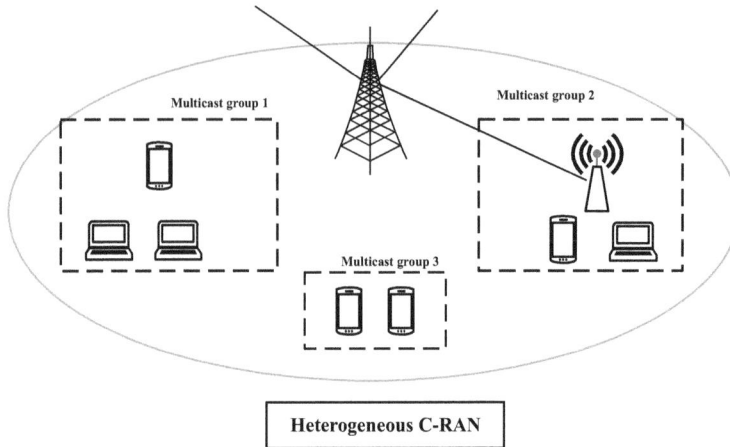

Fig. (15). Beamforming technology for high-speed data transmission.

The first kind of synthesis problem is based on the pre-given index of the main lobe width and sidelobe level of the pattern (there are no strict requirements for other details of the other pattern). To determine the four factors of the array pattern (the number of array elements, the array spacing, the excitation amplitude of the array elements, and some of the excitation phase of the array elements. The most famous methods of this kind of synthesis are Chebyshev synthesis, Taylor synthesis, and so on.

The second kind of synthesis problem is the synthesis of specifying the shape of the desired pattern. This kind of method is a problem of function approximation. The commonly used methods are the interpolation method, Woodward-Lawson synthesis method, optimization calculation method, and so on.

The third kind of synthesis problem is the synthesis of perturbation of array spacing and excitation amplitude. The fourth kind of synthesis problem is to optimize a specific parameter of the array antenna (such as the width of the main lobe) to obtain the weight that meets the requirements of the given pattern.

For large-scale antennas, the goal of array design needs to meet the comprehensive requirements of mobile communication, such as common channel and service repair coverage, spectrum efficiency improvement, and so on. Under the condition of area constraints, the performance is optimized as much as possible, and the implementation complexity is reduced.

The radiation characteristics of array antennas depend on element factors and array factors. The element factors include the amplitude and phase of the exciting current, the voltage standing wave ratio, the gain, the pattern, and the polarization mode of the array element. The array factors mainly include the number of array elements, the arrangement of array elements, and the spacing of array elements. Although the radiation characteristics can be changed by controlling the array factors, the element factors also need to be paid attention to when controlling the overall characteristics of the array. The large-scale antenna array is generally a rectangular plane array, and the array elements are arranged in the plane according to the rectangular grid, as shown in Fig. (**16**). The array is distributed in the XOY plane with 2Mx row array elements spaced by dx. It is distributed in the X-axis direction, 2Ny, and the array elements are distributed in the Y-axis direction with the spacing dy.

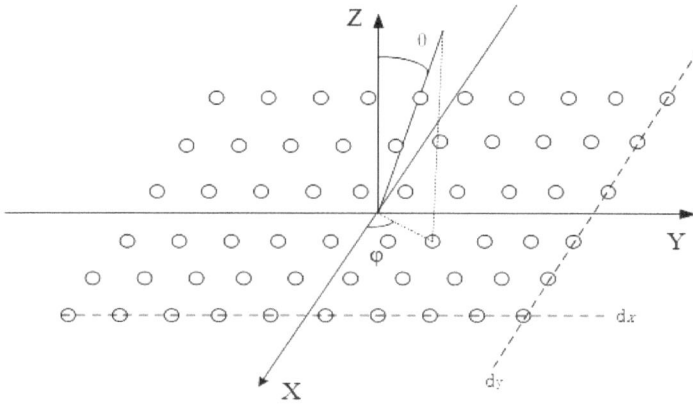

Fig. (16). Schematic diagram of the large-scale rectangular plane array.

Assuming that all elements are isotropic, (m, n) is the m-th row and n-th column of the two-dimensional matrix $(2M_x, 2N_y)$, and (ξ_m, η_n) represents the position of the (m, n) element, then $\xi_m = md_x, \eta_n = nd_y$, where $-N_X \leq m \leq N_x$; N_x . If the current amplitude and phase distribution of the (m, n) cell are expressed by I_{mn} and Φ_{mn}, the array factor of the rectangular planar array can be written as

$$F(\theta,\varphi) = \sum_{m=-N_x}^{N_x} \sum_{n=-N_y}^{N_y} \{(\frac{I_{mn}}{I_{00}}) \exp(j(\varphi_{mn} - \varphi_{00})$$
$$\exp[jk \sin\theta(\xi_m \cos\phi_{mn} + \eta_n \sin\phi_{mn})]\}$$

(42)

From the above Equation (**42**), the matrix factor of the rectangular matrix is equal to the product of the two linear matrix factors in the x-axis direction and the y-axis direction, so the one-dimensional linear array analysis method can be used to repeatedly analyze the two-dimensional rectangular area array.

The pattern of the array antenna has a specific beam scanning function, called beam deflection, which can scan the maximum angle of antenna gain by controlling the amplitude and phase of the feed of each element. For two adjacent elements $N(i)$ and $N(i+1)$, to achieve the maximum scanning angle, the feed phase of the element $N(i+1)$ needs to be ahead of the feed phase of the element to compensate for the spatial wave path difference. If another place ϕ is satisfied the Equation **(42)**, which creates a gate valve in the direction of ϕ mailbox mechanism similar to that of the main lobe, therefore cannot be eliminated by array weight optimization, concrete as shown Fig. **(17)**.

$$2\pi \frac{d\sin\theta + d\sin\phi}{\lambda} = 2n\pi \qquad n = 1, 2, \cdots N \qquad (43)$$

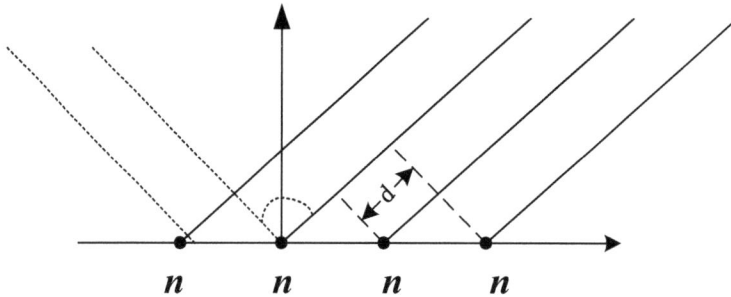

Fig. (17). Principle of beam forming for line array.

For the array element spacing d beam scanning within, the condition that no grating lobes appear in the pattern is satisfied the Equation **(44)**. Where θ_{max} is the maximum scanning angle, the constraint condition for the distance between the array elements can be obtained is $d \ll \lambda/2$.

$$d < \frac{\lambda}{1 + |\sin\theta_{msx}|} \qquad (44)$$

The array antenna can be regarded as an aperture antenna. Therefore, the aperture size and electric field distribution restrict its direction. Generally, when the size of the array antenna is determined, constant amplitude and in-phase excitation can make the array antenna obtain the strongest directivity, that is, the narrowest beam. The minimum side boundary condition of the broadcast beamwidth range can be expressed as:

$$\theta_{\min} = 0.886 \times \frac{\lambda}{N_d} \qquad (45)$$

In the above equation, the Angle of coverage of the narrowest beam width is. The number of the antenna array is N. This equation can be used as the minimum boundary condition of the broadcast beam width range. As for the maximum boundary, it is not restricted by the aperture size, so there is no definite width constraint.

Null-notch Beamforming Algorithm Based on LMS Criterion

The current research on multi-user beamforming algorithms can be divided into two categories, traditional beamforming algorithms that reduce multi-user interference and traditional beamforming algorithms that add nulling to completely eliminate multi-user interference [10]. Typical traditional beamforming algorithms to reduce multi-user interference include the ZF beamforming algorithm, Max-SNR beamforming algorithm, and MMSE beamforming algorithm. Nowadays, the ZF beamforming algorithm and MMSE beamforming algorithm are more commonly used. The main purpose of this type of algorithm is to maximize the transmission power and suppress multi-user interference to a certain extent. However, due to the low degree of multi-user interference reduction in this type of algorithm, or the sidelobe aligning the interference direction to re-introduce interference, the use of this type of algorithm will cause some inter-user interference in scenarios where there are many users in space division multiplexing., This part of the inter-user interference will affect the system performance. The best way to solve this kind of problem is to set the null in the specified direction. Therefore, people have researched setting null-notch beamforming algorithm.

Null-notch beamforming algorithms can be classified into wide-null-notch beamforming algorithms and sharp-null-notch beamforming algorithms according to the shape of the null-notch. They can also be classified into null beamforming algorithm based on LMS criterion, null beamforming algorithm based on genetic algorithm and null beamforming algorithm based on phase or position disturbance according to the criteria or algorithms used. The purpose of these algorithms is the same, they are all forming a directional narrow beam in the desired direction, and forming a null in the interference direction to suppress multi-user interference. The system capacity can be effectively improved. However, the existing null-notch beamforming algorithms will cause different degrees of loss in the gain of the main lobe. The main reason is that compared with the traditional beamforming algorithm, the nulling beamforming algorithm increases the constraint condition of setting nulling, which limits the freedom of the algorithm, and finally leads to the reduction of the radiation power of the beam in the desired direction. In the beamforming algorithm with nulling, the more nulls are set, the lower the radiation power of the beam in the desired direction. Besides, the setting of strong constraints can also cause beam distortion. Therefore, how to design an algorithm to minimize the beam radiation power loss in the desired direction and maximize the null depth in the interference direction is one of the problems that need to be solved today.

According to the Massive MIMO system model, the form of the transmitted signal of the antenna array can be expressed as:

$$y(n) = w(n)^H x(n) \tag{46}$$

Among them, $w(n)$ represents the antenna array weight vector, $x(n)$ represents the transmitted signal data, assuming that the expected array transmission signal of the null-notch beamforming algorithm based on the LMS criterion is $d(n)$, the expected antenna array transmission can be calculated The error between the signal and the actual antenna array transmitted signal is:

$$e(n) = d(n) - y(n) = d(n) - w(n)^H x(n) \tag{47}$$

The mean square error between the expected antenna array transmit signal and the actual antenna array transmit signal is

$$\xi = E[|e(n)^2|] = E[|(d(n) - w(n)^H x(n))^2|] \tag{48}$$

The LMS criterion is similar to the MMSE criterion. Both are to obtain the weight vector $w(n)$ when ξ takes the minimum value. The MMSE criterion uses the method of derivation and zeroing to solve the optimal beamforming weight vector, while the LMS criterion the optimal beamforming weight vector is obtained by the recursive method. The recursive update direction is the decreasing direction of ξ, that is, the gradient direction of ξ. The recursive formula under the LMS criterion is:

$$w(n+1) = w(n) - \kappa \frac{\partial E[|e(n)^2|]}{\partial w(n)} \quad w(n+1) = w(n) - \kappa \frac{\partial E[|e(n)^2|]}{\partial w(n)} \tag{49}$$

The derivation of ξ can be obtained:

$$\frac{\partial \xi}{\partial w(n)} = \frac{\partial E\left[|e(n)^2|\right]}{\partial w(n)} = 2e^H(n)\frac{\partial e(n)}{\partial w(n)} = 2e^H(n)x(n) \qquad (50)$$

It can be obtained by combining Equation (**49**) and Equation (**50**)

$$w(n+1) = w(n) - \kappa 2e^H(n)x(n) = w(n) - \mu e^H(n)x(n) \qquad (51)$$

where u is the recursive update step and needs to meet:

$$0 < \mu < (ME\{x(n)^2\}) \qquad (52)$$

Assuming that the initial value of the beamforming weighting vector is $w(0)=[0,0,...,0]$, the length of the $w(n)$ weighting vector is M, and the optimal beamforming weighting vector can be solved according to the steps in the basic flow of the LMS algorithm.

The computational complexity of the LMS algorithm is relatively low, although it needs to be solved recursively, there is only one item in each recursion, with N multiplication and addition at most, and N is the number of antennas. Moreover, the LMS algorithm starts from the signal domain and can form effective zeros on the beam. However, the power spectrum of the weighted signal based on the LMS zero notch beamforming algorithm is relatively steep, which causes a lot of trouble for the fabrication of the power amplifier at the signal transmitter, so it is indispensable to study the improved algorithm, making the power spectrum flatter. The improved LMS zero-notch beamforming algorithms mainly include the zero-notch beamforming algorithm based on the normalized least mean square (Normalized Least Mean Square, NLMS) criterion and the zero-notch beamforming algorithm based on the symbolic least mean square (Sign Least Mean Square, SLMS) algorithm.

As shown in Table **2**, the null-notch beamforming algorithm based on the NLMS criterion mainly changes the recursive update direction. Compared with the recursive update direction of the LMS, the null-notch beamforming algorithm under the NLMS criterion is a return One operation. Therefore, the recursive formula of the null-notch beamforming algorithm based on the NLMS criterion is

Table 2. The basic flow of the LMS algorithm.

1: Determine the recursion length M and recursive update step size pa.
2: Determine the initial beamforming weight vector w(o).
3: Construct the desired transmission signal d(n) according to the desired null setting.
4: Construct the transmission signal x(n) according to the transmission signal form.
5: Perform recursion to solve w(n).

$$for n = 1:M$$
$$y(n) = w(n)^H x(n)$$
$$e(n) = d(n) - y(n)$$
$$w(n+1) = w(n) + \mu x(n)e^H(n)$$
$$end$$

6: The beamforming weighting vector is obtained and the signal is weighted.

$$w(n+1) = w(n) + \frac{\mu}{\|x(n)\|} x(n)e^H(n) \tag{53}$$

In the high-speed communication scenario, the running speed of the algorithm needs to be further improved. At this time, only the polarity of the error function

can be considered, so the zero-notch beamforming algorithm based on SLMS criterion is produced. This algorithm can significantly reduce the complexity of the algorithm, and its recurrence formula can be expressed as:

$$w(n+1) = w(n) + \mu x(n)e^H(n)\operatorname{sgn}[x(n)] \tag{54}$$

Among them, $\operatorname{sgn}\left[x(n)\right]$ is a symbolic function with the argument $x(n)$, which can be expressed as:

$$\operatorname{sgn}[x(n)] = \begin{cases} 1 & x(n) > 0 \\ 0 & x(n) = 0 \\ -1 & x(n) < 0 \end{cases} \tag{55}$$

The null-notch beamforming algorithm based on the LMS criterion can generate beams with accurate nulls, which reduces the multi-user interference of the system.

DESIGN OF FEED NETWORK AND RF FRONT-END

Feeding Technology of Base Station Antenna

The feed network is an important component of the base station antenna, which connects the antenna port and the array elements to form the path of RF signal transmission, and realizes the functions of impedance matching, amplitude and phase distribution, and so on. It is closely related to the performance of antenna array, which directly affects the standing wave ratio, radiation efficiency, beam direction, sidelobe level, zero, and other parameters of the antenna array [11]. In the design of the feed network of the base station antenna, we focus on the impedance matching, insertion loss, amplitude, and phase distribution of the feed network, to reduce the loss of the antenna array, improve the radiation efficiency and antenna gain. Good sidelobe suppression and other pattern characteristics are obtained. To realize the broadband base station antenna, the plate antennas used by the base station use parallel feeding. The basic radiation unit of the base station antenna is commonly used as the long dipole antenna. There are two kinds of input impedance, most of which are 50 Ω and a few are 75 Ω. Because there are only 50 Ω and 75 Ω coaxial lines, the impedance transformation section can only combine them in parallel to achieve the desired characteristic impedance.

The eight-unit plate antenna feed network is shown in Fig. (**18**). Basic unit to shoot the input impedance of 50 Ω plate-like antenna feed network. To ensure equal amplitude and same direction feeding, eight equal lengths 50Ω coaxial lines ① ~ ⑧ are directly connected with 1~8 long dipoles and connected in parallel. Parallel impedance is 50/2 = 25Ω.Using impedance transformation section ⑨ ~ ⑫ of odd times 75 T coaxial line, 25 Ω is changed into 225 Ω (75²/25 = 225), and 225/2 = 112.5 Ω after being connected in parallel.

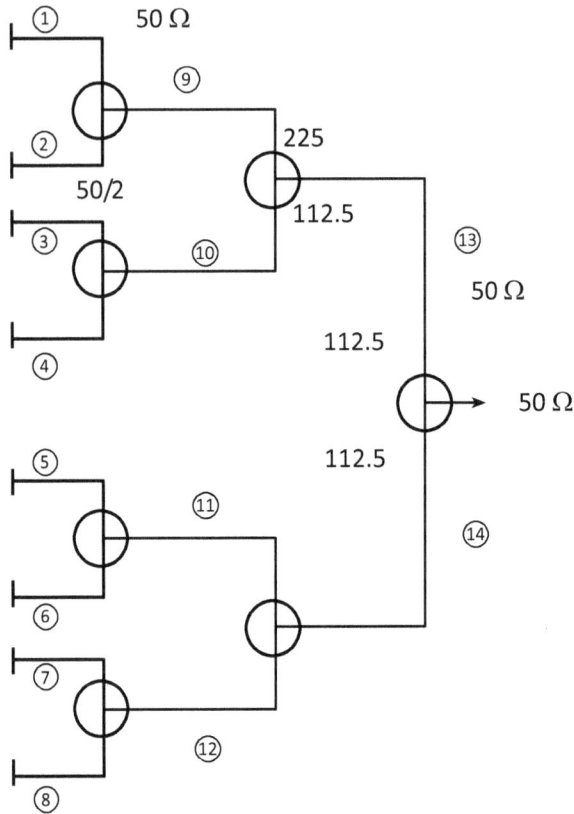

Fig. (18). Eight-unit plate antenna feed network.

Eight-element patch antenna array can be fed in parallel or phase, as shown in Fig. **19** (**a** and **b**). In Fig. (**19a**), the radiating elements of the antenna are divided into two groups, which are fed in parallel in anti-phase, and then the lengths of the two feeder lines are different by, which becomes in phase. It can also be regarded as that two adjacent radiating elements are fed in parallel with opposite phases by two power splitters, many of which are located on both sides of the patch antenna. Fig. (**19b**) shows that many two-power splitters located on the same side of the patch antenna are used to feed adjacent radiating elements in phase and parallel.

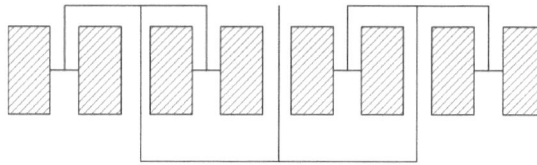

(a)

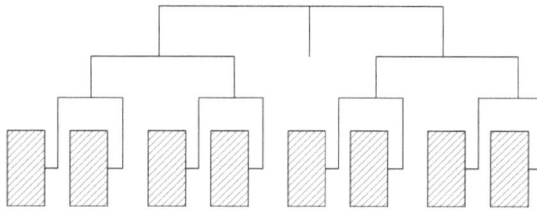

(b)

Fig. (19). Sketch of eight-element patch antenna array.

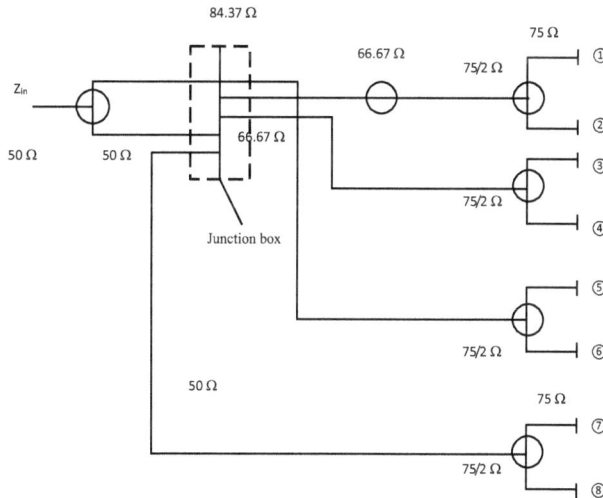

Fig. (20). Sketch of eight-unit plate antenna feeder network.

Compared with the in-phase parallel feed network, the inverted feed network has a symmetrical structure, and occupies a relatively small space when the line array forms a planar array, and is easy to suppress cross-polarization components. Similarly, we take the eight-unit plate antenna feeder network as an example. The eight-element antenna feeding network with a basic input impedance of 75Ω is shown in Fig. (**20**).

Connected with the corresponding dipole by eight equal lengths 75Ω coaxial lines, And in parallel, Parallel impedance: 75/2=37.5Ω. The impedance transformation section 37.5Ω into 66.67Ω; by using the coaxial line with an odd number of times length 50 Using the impedance transformation section of the coaxial line 75Ω long to change 66.67 to 84.37Ω; A coaxial line with impedance 84.37 Ω are connected in parallel with a coaxial line with impedance 66.67 Ω through a junction box, A parallel impedance of 17.59 Ω; Use one length 50 Ω coaxial line and one length 30 Ω. The characteristics of 17.59 Ω parallel are transformed into 51.2Ω by impedance transformation section with resistance to 50 Ω.

Design of RF Front-End for Large-Scale Active Antenna

With the advent of the 6G era, the design of the RF front-end becomes more complex. The new wireless network requires more RF front-end functions, including high-order MIMO and Massive MIMO, smart antenna systems, and complex filtering functions. The higher the MIMO order, the more antennas are needed, the more antennas are needed, the more feeder lines are needed, and the more feeder interfaces are used in RRU, so the complexity of the process becomes higher and higher. Besides, there is a certain loss of the feeder itself, which will also affect the performance of part of the system. For this reason, some of the physical layer processing functions of BBU will be merged with the original RRU into AAU (active antenna processing unit) in the 6G system.

Taking the large-scale active antenna architecture (BBU+AAU) as an example, this section introduces the RF front-end design of large-scale active antennas, which mainly includes two parts: RF part design and digital if the design. Because the planned AAU supports 128 RF channels, if 128 RF channels are realized on a direct veneer, the size of the board is very large, and the process is difficult to implement. To ensure the realizability and manufacturability, the overall architecture of AAU adopts a modular design, which divides the 128 active channels into four groups of 32-channel active systems, thus greatly reducing the size of a single board and ensuring the manufacturability of the system, as shown in Fig. (**21**).

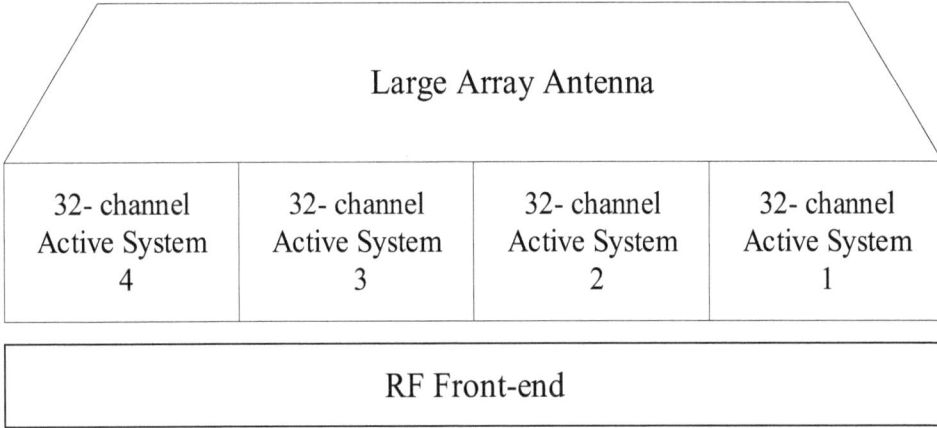

Fig. (21). Architecture of large-scale active antenna.

The large-scale antenna array combines the beard connector with the RF front-end to form a complete large-scale active antenna system [12]. Among them, each group of the 32-channel active system includes a motherboard, a power amplifier low noise amplifier board, and a power board, as shown in Fig. (**22**).

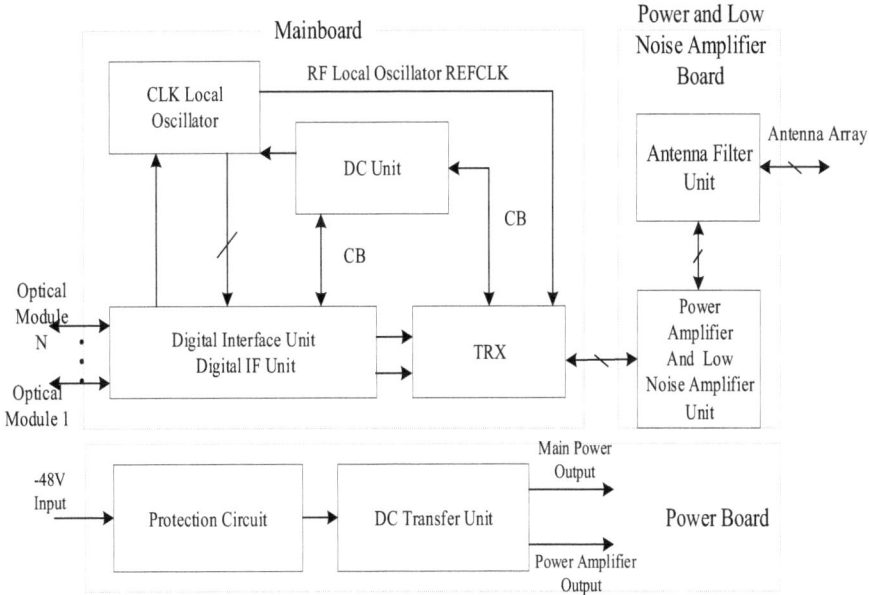

Fig. (22). Sketch of 32-channel active system.

The power amplifier low noise amplifier template carries a power amplification unit, low noise amplification unit, and antenna filter unit, which completes the power amplification of 32 downlink signals, feedback of output signals, low noise amplification of 32 uplink signals, and auxiliary functions such as calibration data storage of each radio frequency channel, which is connected to the antenna array through radio frequency connectors. The motherboard carries a digital control unit, data interface unit, digital intermediate frequency unit, system clock local oscillator unit, and transceiver unit. It realizes interface connection with the master station, base-band remote protocol, base-band IQ data distribution, timing control, main control processing, 32-channel digital intermediate frequency processing, DDC (Digital Down Converter), DUC (Digital Up-Converter), CFR (Crest Factor Reduction), DPD (Digital Pre-Distortion), local oscillator generation and distribution, DPD feedback channel, 32-channel transceiver processing. The power supply board completes the power port protection and filter and generates the working voltage required by the motherboard and power amplifier low-noise amplifier card by isolating the power converter.

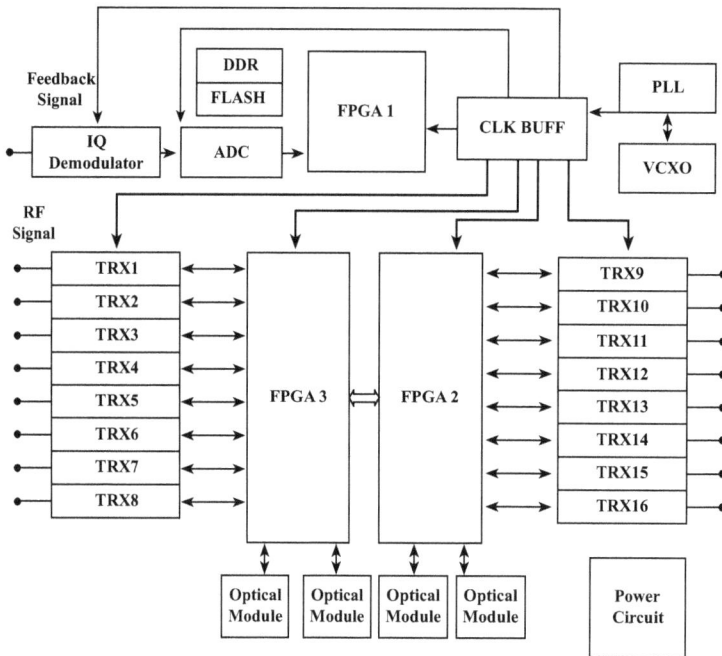

Fig. (23). Implementation diagram of the motherboard.

The implementation diagram of the motherboard is shown in Fig. (**23**). The motherboard completes the control, IR interface (the interface between BBU and RRU), digital intermediate frequency, and small-signal analog transceiver. The processor color uses an FPGA with ARM (Field-programmable Gate Array), IR interface, and digital intermediate frequency are realized by large capacity FPGA. The analog small-signal transceiver adopts an integrated TRX (Transceiver), feedback channel and uses the separate channel to realize the feedback loop. The main functions of the motherboard include the following.

The implementation of the baseband stretch protocol is to communicate with the indoor unit of the base station mainly through the optical fiber transmission link, receive and send I/Q baseband signals, and the corresponding OSM (Operation and Maintenance) information.

According to the control information of the main carrier, the time slot control signal is generated, and the time slot control signal, switch signal and switch matrix meeting the requirements of transceiver are provided.

The multi-antenna signal to be sent is distributed to each digital-to-analog conversion to become an intermediate frequency signal, which is output after upconversion and amplification.

After analog downconversion, analog-to-digital conversion, and downconversion processing, the RF signal is transformed into a multi-antenna baseband signal, which is processed by AGC (automatic gain control) and sent to the corresponding base station indoor unit through the optical fiber according to the pull-out protocol.

The function of local clock recovery provides a highly stable clock signal through the local frequency synthesizer, while filtering out the noise introduced in the transmission and reducing the jitter of the clock signal.

Generally speaking, the transceiver can adopt two schemes: discrete or integrated. The discrete scheme is composed of a modulator, mixer, wood oscillator, amplifier, digital-to-analog conversion (ADC), analog-to-digital conversion (DAC), *etc*. The transmission link is composed of DAC, quadrature modulator, LC matching circuit, and RF digital variable gain amplifier (DVGA). The gain change caused by temperature change and board inconsistency is adjusted by RFDVGA. The receiving link consists of ADC, mixer if DVGA, and SAW Surface Acoustic Wave filter. The feedback link consists of a demodulator and a feedback ADC. The calibration channel multiplexes the transceiver channel. The integrated scheme adopts an integrated transceiver chip, and the link is simplified to an integrated transceiver, amplifier, and filter. The number of AAU channels is large, and the size of the board is small. To realize multi-channel integration, the integration scheme is adopted. The transceiver, amplifier, and filter are integrated to form the transceiver link to realize the functions of analog signal amplification, filtering, frequency conversion, and gain adjustment.

The power amplifying unit amplifies the downlink radio frequency signal and realizes the coverage of the downlink signal. The power amplifier module comprises a power amplifier pre-drive tube, drive tube, final amplifier tube ring coupler, and other devices. A single power amplifier achieves 3.4~3.6GHz band, signal bandwidth ≥ 200MHz, average output power ≥ 3W, power amplifier efficiency ≥ 40%, and downlink ACLR (Adjacent Channel Leakage Ratio) is better than-45dBc. The schematic diagram of the TX transmission link is shown in Fig. (**24**).

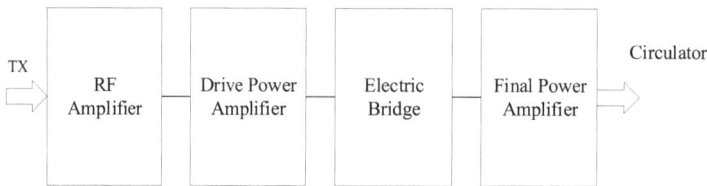

Fig. (24). Schematic diagram of the TX transmission link.

Due to the high frequency of 3.5 GHz and the need to support 200MHz RF bandwidth, the traditional LDMOS power amplifier has low efficiency at this frequency point, because the power amplifier chooses GaN power amplifier to complete the design, and realizes the broadband design while meeting the high efficiency.

ANTENNA SELECTION TECHNOLOGY

Antenna Selection Criteria and Classification

MIMO and Massive MIMO technology obtain the degree of freedom of the spatial dimension by increasing the number of antennas, significantly improving the system capacity, improving the signal quality and increasing the reliability of the communication link without the increase of the spectrum resources. However, in the actual commercial application deployment, the base station not only needs to consider the system performance but also the cost is a very essential consideration. In a wireless mobile communication system, the cost proportion of the RF transceiver link is usually relatively high [13]. With the increase of the number of antennas, especially the number of antennas in Massive MIMO, the number of RF links increases greatly, which makes the hardware cost and maintenance cost increase considerably, and the complexity of system hardware implementation will also increase significantly. Besides, with the substantial increase in the number of antennas, the possibility of some antennas corresponding to deep fading wireless channels also increases, so it is not cost-effective to pay high RF hardware link cost for such channels.

The birth of Antenna selection (AS) technology is to solve this contradiction. Antenna selection technology is also commonly known as antenna subset selection technology (Antenna Subset Selection) [14]. It refers to the process that the optimal antenna subset is selected from all antennas according to certain criteria and allocated to limited RF link resources when the number of antennas installed at the transmitter or receiver is more than the number of RF links. The schematic diagram of a multi-antenna system with antenna selection function is shown in Fig. (25). Antenna selection technology not only effectively reduces the hardware cost and hardware implementation complexity of the Massive MIMO system but also maximizes some technical features and system gain of the Massive MIMO system.

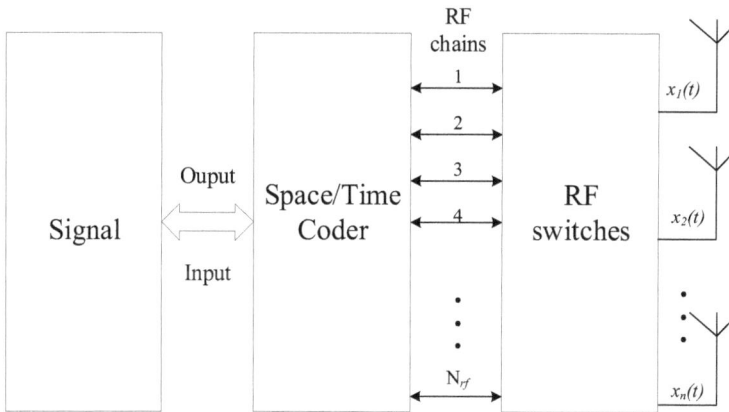

Fig. (25). Schematic diagram of a multi-antenna system with antenna selection function.

Antenna selection technology is divided according to the selection criteria, including channel capacity maximization criterion, based on signal-to-noise ratio maximization criterion, bit error rate minimization criterion, energy efficiency maximization criterion, *etc.* According to the application objects of antenna selection algorithm, it can be divided into transmitting antenna selection, receiving antenna selection, and joint transceiver antenna selection. According to the performance of the antenna selection algorithm, it can be divided into the optimal algorithm and suboptimal algorithm. The optimal antenna selection algorithm usually refers to the exhaustive search algorithm, but the computational complexity of the exhaustive search algorithm is very high and can only be used as a theoretical reference, especially in the Massive MIMO system with a large number of antennas, the calculation results are difficult to get, even cannot be obtained. Therefore, most of the literature on antenna selection are devoted to the research of suboptimal antenna selection algorithms with low complexity, such as classical incremental antenna selection algorithm, decreasing antenna selection algorithm, channel matrix row or column norm maximization algorithm, and so on.

Assuming that Q antennas are used in the NT antennas, the effective channel coefficient matrix can be described by the Q column in H. Let P$_i$, be the sequence number corresponding to the I selected column vector in H, where, so the corresponding $NR{\times}Q$ dimensional effective channel coefficient matrix can be expressed as the. Let be a space-time coded transmitted signal or a spatially multiplexed stream transmitted signal mapped to Q antennas, so the received signal can be expressed as:

$$y = H_{\{P_1,P_2,\cdots P_Q\}}x + Z \tag{56}$$

where the $z \in C^{N_R\times1}$ is the additive white Gaussian noise. The channel capacity of the system described in Equation **(56)** depends on the selected subset of antennas and the number of selected antennas.

Optimal Antenna Selection Algorithm

By using the method of complete enumeration, that is, exhaustive search, the optimal antenna algorithm calculates the channel capacity of all possible antenna combinations and selects the antenna subset with the largest channel capacity. Q antennas are selected from N antennas to maximize the channel capacity of the system. if the total transmission power is fixed, the channel capacity of the system can be expressed as:

$$C = \max_{H_{\{p_1,p_2,\cdots p_Q\}} \in H} \log_2 \det\left(I_{N_R} + \frac{\rho}{Q} H_{\{p_1,p_2,\cdots p_Q\}} R_{xx} H^H{}_{\{p_1,p_2,\cdots p_Q\}} \right) \tag{57}$$

Among them, Rxx is a $Q{\times}Q$ dimensional covariance matrix. If the transmitter has equal power distribution, there are

$$C_{\{p_1,p_2,\cdots p_Q\}} = \log_2 \det\left(I_{N_R} + \frac{\rho}{Q} H_{\{p_1,p_2,\cdots p_Q\}} R_{xx} H^H{}_{\{p_1,p_2,\cdots p_Q\}} \right) \tag{58}$$

To obtain an antenna subset with Q antenna optimum, all possible antenna combinations need to be calculated according to Equation **(58)**. To maximize capacity performance, the selected antenna subset must maximize the Equation **(58)**, that is

$$\left\{p_1, p_2 \cdots p_Q\right\}^{opt} = \arg \max_{\left\{p_1, p_2 \cdots p_Q\right\} \subset \mathbb{S}_Q} C_{\left\{p_1, p_2 \cdots p_Q\right\}} \tag{59}$$

where represents the collection of all antenna combinations, $\left|\mathbb{S}_Q\right| = C_{N_T}^Q$. As can be seen, if Equation **(59)** solves all antenna combinations, the computation is very large and is bound to be very time-consuming. This is unacceptable for mobile communication systems requiring low delay, especially when the *NT* is very large.

Incremental Antenna Selection Algorithm

The incremental antenna selection algorithm is a suboptimal antenna selection method, which is arranged in ascending order through channel capacity, and then selects the antenna corresponding to the maximum channel capacity until the specified antenna selection number is reached. Especially, the first option should be to enable the system to obtain maximum channel capacity antenna.

$$
\begin{aligned}
p_1^{subopt} &= \arg \max_{p_1} C_{\{p_1\}} \\
&= \arg \max_{p_1} \log 2 \det\left(I_{N_R} + \frac{\rho}{Q} H_{\{p_1\}} H^H{}_{\{p_1\}} \right)
\end{aligned}
\tag{60}
$$

After completing the selection of the first antenna, select the second antenna in the following way to maximize the system capacity.

$$
\begin{aligned}
p_2^{subopt} &= \arg \max_{p_2 \neq p_1^{subopt}} C_{\{p_1\}} \\
&= \arg \max_{p_2 \neq p_1^{subopt}} \log_2 \det\left(I_{N_R} + \frac{\rho}{Q} H_{\left\{p_1^{subopt}, p_2\right\}} H^H{}_{\left\{p_1^{subopt}, p_2\right\}} \right)
\end{aligned}
\tag{61}
$$

Suppose you get after the n selection. For the $n+1$ antenna selection, the channel capacity can be calculated under

$$
\begin{aligned}
C_l &= \log_2 \det\left(I_{N_R} + \frac{\rho}{Q}(H_{\{p_1{}^{subopt},p_2{}^{subopt},\cdots p_n{}^{subopt}\}} H^H{}_{\{p_1{}^{subopt},p_2{}^{subopt},\cdots p_n{}^{subopt}\}} + H_{\{l\}} H^H{}_{\{l\}}) \right) \\
&= \log_2 \det\left(I_{N_R} + \frac{\rho}{Q}(H_{\{p_1{}^{subopt},p_2{}^{subopt},\cdots p_n{}^{subopt}\}} H^H{}_{\{p_1{}^{subopt},p_2{}^{subopt},\cdots p_n{}^{subopt}\}} \right) \\
&= \log_2 \det\left(1 + \frac{\rho}{Q}H_l(I_{N_R} + \frac{\rho}{Q}(H_{\{p_1{}^{subopt},p_2{}^{subopt},\cdots p_n{}^{subopt}\}} H^H{}_{\{p_1{}^{subopt},p_2{}^{subopt},\cdots p_n{}^{subopt}\}})^{-1} H^H{}_{\{l\}}) \right)
\end{aligned}
\tag{62}
$$

The $(n+1)$th selected antenna needs to be made. To maximize, we have

$$
\begin{aligned}
p_{n+1}{}^{subopt} &= \underset{l \notin \left\{p_1{}^{subopt},p_2{}^{subopt},\cdots p_n{}^{subopt}\right\}}{\arg\max} C_l \\
&= {}_{l \in \{1,2\cdots N_T\}-\left\{p_1{}^{subopt},p_2{}^{subopt},\cdots p_n{}^{subopt}\right\}} H_{\{l\}} \\
&\left(\frac{Q}{\rho}I_{N_R} + H_{\{p_1{}^{subopt},p_2{}^{subopt},\cdots p_n{}^{subopt}\}} H^H{}_{\{p_1{}^{subopt},p_2{}^{subopt},\cdots p_n{}^{subopt}\}}\right)^{-1} H^H{}_l
\end{aligned}
\tag{63}
$$

Calculate the Equation **(63)** continuously until all Q antennas are selected. During the whole antenna selection process, for all, $l \in \{1,2\cdots N_T\}-\left\{p_1{}^{subopt},p_2{}^{subopt},\cdots p_n{}^{subopt}\right\}$, only one matrix reverse operation is needed.

Decreasing Antenna Selection Algorithm

The decreasing antenna selection algorithm is also a suboptimal algorithm. By descending the channel capacity, the antenna corresponding to the minimum channel capacity is eliminated until the remaining antenna is exactly equal to the specified antenna selection number. Obviously, decreasing the antenna selection algorithm is opposite to that of the increasing antenna selection algorithm. When the, is selected as the n antenna serial number set, in the initial state, S = {1, 2 ... N_T}, and the antenna with the least contribution to the channel capacity of the system is selected, we have

$$p_1^{deleted} = \frac{\arg\max}{p_1 \in \mathbb{S}_1} \log_2 \det\left(I_{N_R} + \frac{Q}{\rho} H_{\mathbb{S}_1-\{p_1\}} \overset{H}{H}^H_{\mathbb{S}_1-\{p_1\}}\right) \tag{64}$$

According to Equation **(64)**, the selected antenna will be removed from the set of antenna serial numbers, and the new set of antenna serial numbers will be

obtained $\mathbb{S}_2 = \mathbb{S}_1 - \{p_1^{deleted}\}$. The antenna, if $|\mathbb{S}_2| = N_T - 1 > Q$ continues to be selected and deleted, and the antenna that needs to be eliminated should be the antenna with the least contribution to the channel capacity in the current antenna set, $\mathbb{S}_2$, we have

$$p_2^{deleted} = \frac{\arg\max}{p_2 \in \mathbb{S}_2} \log_2 \det\left(I_{N_R} + \frac{Q}{\rho} H_{\mathbb{S}_2-\{p_2\}} H^H_{\mathbb{S}_2-\{p_2\}}\right) \tag{65}$$

Therefore, the new set of antenna serial numbers is updated to that the Equation (65) need to be repeated until all $N_T - Q$ antennas are selected and eliminated $|\mathbb{S}_m| = Q$.

The incremental and decreasing antenna selection algorithm has a similar principle and process. The former selects and adds the antenna which contributes the most to the channel capacity of the system one by one, but only considers the influence of a single antenna on the system. In particular, the first antenna selection ignores the influence of each antenna; the latter selects and deletes one by one from the full set.

Each selection process takes into account the influence of the remaining antenna on the system. From the point of view of complexity, the decrement selection algorithm is higher than the incremental selection algorithm, but when $1<Q<N$, the capacity performance of the decrement selection algorithm is higher than that of the incremental selection algorithm. This is because the decrement selection algorithm takes into account the correlation between all column vectors of the initial channel coefficient matrix when selecting the first deleted antenna. As a whole, the antenna selection algorithms of increment and decrement are suboptimal, except for the following two cases: the decreasing algorithm and the Equation **(65)**. The antenna subset obtained by the optimal antenna selection

described is the same; when $Q=1$, the incremental algorithm is the same as the antenna subset obtained by the optimal antenna selection represented by Equation **(65).**

MEASUREMENT TECHNOLOGY OF MASSIVE MIMO ANTENNA

As for 6G technology, massive access is also a problem to be solved, where it needs to meet the access of a massive number of wireless external connections at the same time and get rid of the interference of bad node information. Differ from the massive MIMO antenna applied in 5G era, the number of base station array antenna elements will increase, even to 10000 in 6G era. Therefore, the measurement technology of Massive MIMO antenna is still required for 6G. Massive MIMO antenna technology construct multi antenna array by increasing the massive number of antennas to compensate the transmission loss of high-frequency path and to improve the spectrum efficiency through spatial multiplexing. Combined with the novel coding technology, it can greatly improve the communication system capacity and communication rate. Therefore, Massive MIMO antenna technology will also be used in 6G mobile communication base stations.

This section will discuss the test and verification of large-scale antenna beamforming technology, including functional verification, performance verification, and so on. A preliminary evaluation of the performance of large-scale antenna beamforming is given. In the past decade, with the research of large-scale antenna beamforming, the development and verification of large-scale antenna prototypes have been carried out in many universities and research institutions. This section introduces the test methods and typical results of the two tests.

OTA Testing Requirements for Massive MIMO Antenna

Over-the-air (OTA) testing is a basic method for communication and radar equipment testing [15]. OTA testing can be carried out on antennas, user equipment, base station equipment, satellite communications, ground stations, radio, and radar, and is suitable for almost all application scenarios. OTA testing is necessary to accurately test complex antenna array systems using beamforming, MIMO and carrier aggregation techniques. The traditional base station equipment is divided into radio frequency unit and antenna unit, and the test is also divided into two parts: 1 radio frequency testing: testing the basic performance of the transmitter and receiver through line explosion connection, such as transmission power, adjacent channel discharge time ratio (ACLR), out-of-band heterozygosity, reception sensitivity, neighbor annoyance selective (ACS) anti-blocking ability, *etc*. For large-scale antennas, since the architecture and working mechanism are different from the traditional antennas, the test scheme is also completely different. Spatial dynamic shaping capability is the most important feature that distinguishes large-scale antennas from traditional antennas. The radiation pattern of the traditional antenna in space is determined, and the antenna can be tested independently. First of all, because the large-scale antenna physically integrates the RF signal excitation with the antenna array, it is impossible to test the antenna array and RF unit separately. at the same time, the function of active signal excitation is as important as the influence of antenna array on beam shape.

OTA measurement parameters can be divided into two categories: research and development, certify cation or conformance testing for the complete evaluation of the radiation characteristics of the equipment under test, and calibration, verification, and functional testing in production. The main test parameters concerned with the design and development of large-scale antennas include gain pattern, radiation power, sensitivity per receiver, transceiver/receiver characteristics, and beam control beam tracking, each of which will expand the OTA measurement. The focus of the test is beam control/beam tracking. Production test conformance and production testing include many aspects, especially the following three aspects. 1. Antenna relative calibration: to achieve accurate beam shaping. The phase integration between the channels of the RF signal must be less than ±5 °, and the measurement can be performed using a phase-coherent receiver to measure the relative error between all antennas.

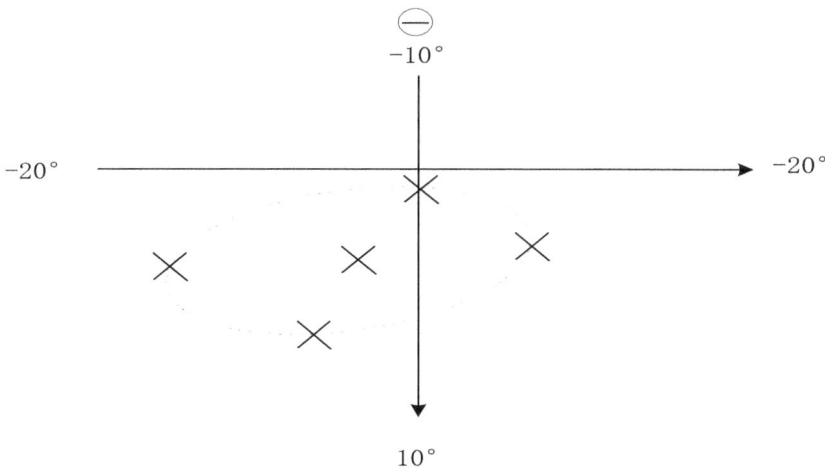

Fig. (26). Sketch of measurement positions for active antenna system.

According to the standard requirements, active antenna system manufacturers are required to specify beam direction, maximum effective Omni-directional radiation power (EIRP, Effective Isotropic Radiated Power), and EIRP threshold for each claimed beam. In addition to the maximum EIRP point, four additional points are measured at the threshold boundary, that is, the center point with the maximum EIRP, and the other four points on the left, right, top, and bottom boundaries, as shown in Fig. (**26**). Final functional testing: performed on fully assembled modules in the production process, including simple radiation testing, 5-point beam testing, and transceiver joint functional testing, such as error vector amplitude measurement (Error vector magnitude) when all transceivers are turned on.

Near-Field and Far-Field Measurement

The OTA measurement system can be classified according to the location of the sampled radiation field. Fig. (**27**) shows the near-field and far-field diagrams of the antenna array from the base station. The near-field and far-field regions are defined by the Fraunhofer distance. In the near field region and the distance which is less than *R*, the field strength is composed of induction component and radiation component, while in the far-field region of the antenna, there is only radiation component field strength.

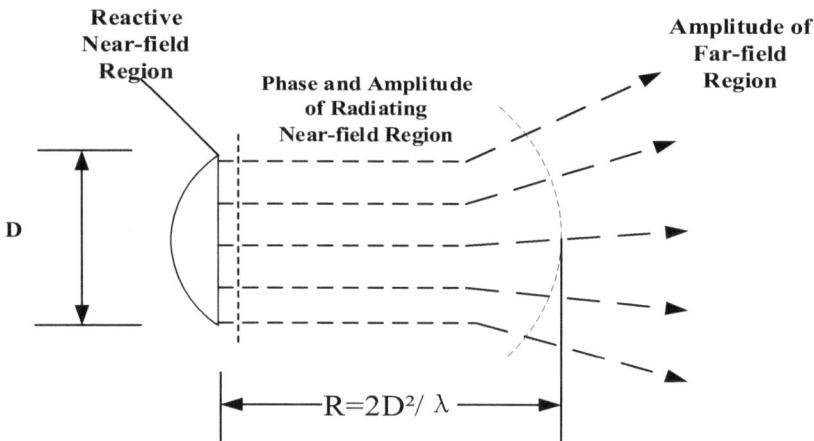

Fig. (27). Principle of OTA measurement system.

Far-Field Test

The method of testing in the far-field area of antenna radiation is called the antenna Far-field Test ($R>2D^2/\lambda$). To measure the far-field area, it is necessary to measure the plane wave amplitude directly, and the anechoic chamber is usually very large, and the size of the darkroom should take into account the size of the equipment to be measured and the measuring frequency. The design of the rectangular anechoic chamber for far-field testing is based on geometrical optics.

According to the properties of the absorbing material, the synthetic field of the reflected waves on the six surfaces of the District Bureau is like the direct radiation of the source antenna to the measuring antenna. In the design, it is usually assumed that the phase change of the reflection point is zero, and the second reflection is not taken into account, and the ultimate goal is to obtain the maximum static region, that is, the simulated far-field area where the antenna is to be left.

Near-Field Test:

The directional encircling test carried out in the near-field area of antenna radiation is called Near-field Test. The antenna near-field measurement system is an automatic measurement system that performs antenna near-field scanning, brain acquisition, test data processing, and test results display and output under the control of the central computer. The whole antenna near-field test system is composed of a hardware subsystem and software subsystem. The hardware subsystem can be further divided into test anechoic chamber subsystem (sampling frame system, multi-axis motion controller, servo driver, near-field test probe, industrial control computer, and peripherals, *etc.*), signal link subsystem (including vector network analyzer system, or time-domain signal source and time-city receiver), data processing computer, *etc.* The software subsystem includes three parts: test control and data acquisition subsystem, data processing subsystem, and result display and output subsystem. In the near-field test, the amplitude and phase distribution of the antenna field is measured at a distance of 3-10λ from the antenna, and the radiation field is calculated by using the stricter mathematical model expansion theory. The near-field test has the advantages of fast measuring speed, small darkroom size and low testing cost. The far-field 3D pattern can be obtained directly by mathematical analysis [16].

The near-field measurement needs to sample the arrival phase and amplitude on the closed surface (spherical, linear, or cylindrical) so that the far-field amplitude can be calculated by the Fourier spectrum transform. For the mathematical transformation to the far-field area, it is necessary to accurately measure the phase and amplitude on the three-dimensional surface of the equipment under test, thus generating the two-dimensional and three-dimensional gain patterns of the antenna. This kind of measurement usually uses a vector network analyzer, with one port connected to the equipment under test and the other port connected to the measuring antenna. Because there are some non-ideal factors in the near-field test, errors will occur in the process of solving the radiation field, and the test accuracy is generally lower than that in the far-field test. The effectiveness of the near-field test needs to be tested by the far-field test.

OTA Testing Process

The basis of all OTA test projects is that the antenna calibration effect is close to ideal, *i.e.*, the amplitude and phase of the output signals of all RF channels are consistent. Therefore, when the test is started, we should follow the order of site calibration, main equipment calibration effect, and OTA test items, and when the effect is not satisfactory in the process of OTA testing, we should recalibrate timely to eliminate the interference caused by poor antenna calibration. In the selection of test sequences in the near field and the far field, the 3D pattern of the equipment to be tested can be measured quickly in the near field, and we can see more details of wave synthesis than the 2D pattern in the far-field. However, due to the near-field active test, it is still in an exploratory period, and the near field test results need to be obtained by post-processing of the algorithm. Compared with the direct test results in the far field, the reliability and stability of the test results are slightly lower. Therefore, if the conditions permit, we can first test from the far-field, and in the OTA test, we first test the receiving power and radiation patterns of a single oscillator, single row, and single row, and compare the test results with the theoretical values to confirm the quality of the calibration effect and the stability of the test site.

The environment of far field testing process is shown in Fig. (**28**). *A* is the equipment to be tested and *D* is the calibrated horn antenna. Before the formal test, we need to calibrate the overall loss of each frequency point of the test field from the receiving instrument to the antenna transmitter of the main equipment.

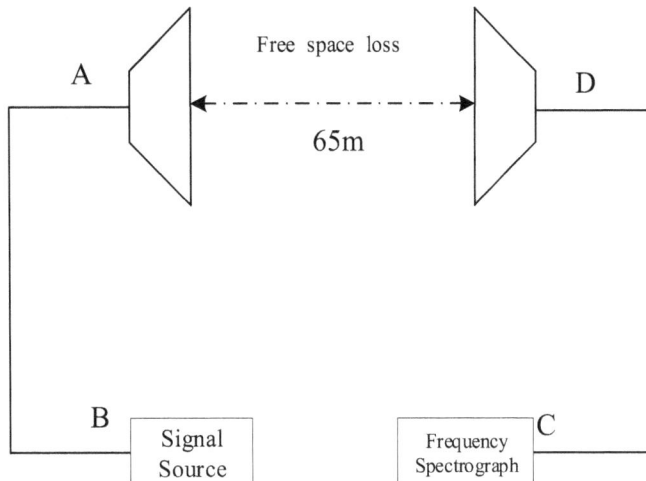

Fig. (28). Far-field testing environment.

The OTA test items carried out in the far-field include effective Omni-directional sensitivity (EIS), EIRR, broadcast wave lake test, service beam, multi-beam test, unit beam test, and so on.

EIS: verifies the uplink receiving radiation sensitivity of large-scale antenna beamforming;

EIRP: verifies various indicators of service beams in different directions under the maximum transmission power of large-scale antennas. The service beam test should cover the direction of the limited coverage pointed by the beamforming in the networking application of the antenna, including the following beams.

Horizontal $0°$, vertical $+3°$ beam.

Horizontal $+60°$, vertical $+3°$ beam.

Horizontal $0°$, vertical $3°+15°$ beam.

Horizontal $+60°$, vertical $3°+10°$ beam.

Horizontal $+10°$, vertical $3°+15°$ beam.

The expected results of the test can be obtained by simulation. Broadcast beam testing: the broadcast beam is mainly used for synchronous channel, broadcast, and pilot signal transmission, and the beam coverage should be consistent with the overall range of desired coverage. Besides, to improve the gain of a broadcast beam, there may be several relatively narrow broadcast beams covering different directions. The broadcast beam needs to give priority to ensuring the gain and half-power beamwidth.

The near field OTA test process generally includes EIRP, unit beam test, broadcast beam test, multi-beam test, and so on. The test steps can be referred to as the far-field test. In the actual testing, the combination of near-field and far-field testing methods are generally used. Functional verification mainly depends on the near-field method to reduce the test cost, while for some strict quantitative performance testing, it needs to be calibrated by far-field testing. When the maturity of the near-field test is high, and the test accuracy is calibrated by the remote test, the follow-up testing can use the near-field environment more.

SUMMARY

This chapter mainly discusses the basic theory of Massive MIMO antenna beamforming technology. The second section introduces the development status of antenna technology in the current wireless communication system. The latest technology development trend of base station antenna and terminal antenna, which are the signal receiving and transmitting ports of wireless access network, is emphasized in terms of miniaturization, integration, multi frequency broadband. And the technical requirements of Massive MIMO technology for large-scale antenna array are summarized. The third section discusses the basic principle of Massive MIMO. This section introduces the basic theory of large-scale antenna array technology in independent and identically distributed Rayleigh fading channel, including the capacity under ideal channel and uplink capacity under pilot pollution. Considering the non-ideal factors for practical application scenarios, the performance of Massive MIMO antenna system is analyzed in depth. The fourth section describes the characteristics and development status of antenna analysis and synthesis technology in Massive MIMO system from the perspective of antenna hardware technology. According to the requirements of beam forming technology, antenna feed network and RF front-end need to be highly integrated with antenna radiation unit to meet the requirements of high-speed communication system. Given the unique requirements of Massive MIMO for antenna performance, the antenna selection method is systematically described and analyzed from the theoretical point of view. Finally, the measurement and correction technology of Massive MIMO antenna suitable for 6G system is introduced, which provides theoretical support for future communication system performance improvement.

CONSENT FOR PUBLICATION

Not applicable.

CONFLICT OF INTEREST

The author declares no conflict of interest, financial or otherwise.

ACKNOWLEDGEMENTS

Declared none.

REFERENCES

[1]   P. Popovski, and W. Connectivity, *An Intuitive and Fundamental Guide.* Wiley Telecom: New York, 2020.

[2]   Y. Zhu, Y. Chen, and S. Yang, "Decoupling and low-profile design of dual-band dual-polarized base station antennas using frequency-selective surface", *IEEE Trans. Antenn. Propag.,* vol. 67, no. 8, pp. 5272-5281, 2019.
[http://dx.doi.org/10.1109/TAP.2019.2916730]

[3]   H.N. Chu, Y-Y. Chen, Y-L. Tsai, and T-G. Ma, "Low-cost polarization sensing system for self-oscillating circularly-polarized active integrated antenna", *IEEE Access,* vol. 7, pp. 170535-170544, 2019.
[http://dx.doi.org/10.1109/ACCESS.2019.2955539]

[4]   S. Kim, J. Choi, and J. Song, "Beam designs for millimeter-wave backhaul with dual-polarized uniform planar arrays", *IEEE Trans. Commun.*, vol. 68, no. 7, pp. 4202-4217, 2020.
[http://dx.doi.org/10.1109/TCOMM.2020.2982382]

[5]   A.B. Sakthi, E.F. Sundarsing, and A. Harshavardhini, "A compact conformal windshield antenna for location tracking on vehicular platforms", *IEEE Trans. Vehicular Technol.*, vol. 68, no. 4, pp. 4202-4217, 2019.

[6]   S.G. Glisic, *Advanced Wireless Networks: Technology and Business Models.* Wiley Telecom: New York, 2016.
[http://dx.doi.org/10.1002/9781119096863]

[7]   Q. Wu, and R. Zhang, "Towards smart and reconfigurable environment: Intelligent reflecting surface aided wireless network", *IEEE Commun. Mag.*, vol. 58, no. 1, pp. 106-112, 2020.
[http://dx.doi.org/10.1109/MCOM.001.1900107]

[8]   Y-M. Zhang, S. Zhang, J-L. Li, and G.F. Pedersen, "A Wavetrap-Based Decoupling Technique for 45° Polarized MIMO Antenna Arrays", *IEEE Trans. Antenn. Propag.*, vol. 68, no. 3, pp. 2148-2157, 2020.
[http://dx.doi.org/10.1109/TAP.2019.2948531]

[9]   H. Asgharimoghaddam, A. Tölli, L. Sanguinetti, and M. Debbah, "Decentralizing Multicell Beamforming via Deterministic Equivalents", *IEEE Trans. Commun.*, vol. 67, no. 3, pp. 1894-1909, 2019.
[http://dx.doi.org/10.1109/TCOMM.2018.2883445]

[10]   H. Li, M. Li, and Q. Liu, "Hybrid beamforming with dynamic subarrays and low-resolution PSs for mmWave MU-MISO systems", *IEEE Trans. Commun.*, vol. 68, no. 1, pp. 602-614, 2020.
[http://dx.doi.org/10.1109/TCOMM.2019.2950905]

[11]   E.A.L. Heinrich, and W. Paul du Plessis, "Numerical optimization of compressive array feed networks", *IEEE Trans. Antenn. Propag.*, vol. 66, no. 7, pp. 3432-3440, 2018.
[http://dx.doi.org/10.1109/TAP.2018.2829834]

[12]   S. Domouchtsidis, C.G. Tsinos, S. Chatzinotas, and B. Ottersten, "Symbol-level precoding for low complexity transmitter architectures in large-scale antenna array systems", *IEEE Trans. Wirel. Commun.*, vol. 18, no. 2, pp. 852-863, 2019.
[http://dx.doi.org/10.1109/TWC.2018.2885525]

[13]   N. Takemura, N. Honma, and A. Kawagoe, "Improvement of interference suppression performance using antenna selection of mobile terminal for Full-Duplex 4×4 MIMO system", *IEEE Trans. Antenn. Propag.*, vol. 68, no. 6, pp. 4186-4195, 2020.
[http://dx.doi.org/10.1109/TAP.2020.2969752]

[14]   M. Hanif, H-C. Yang, G. Boudreau, E. Sich, and H. Seyedmehdi, "Antenna subset selection for massive MIMO systems: A trace-based sequential approach for sum rate maximization", *J. Commun. Netw. (Seoul)*, vol. 20, no. 2, pp. 144-155, 2018.
[http://dx.doi.org/10.1109/JCN.2018.000022]

[15]   P. Kyösti, "Correlation of Hybrid Beamforming Arrays in the Context of OTA Testing", *IEEE Antennas Wirel. Propag. Lett.*, vol. 19, no. 4, pp. 671-675, 2020.
[http://dx.doi.org/10.1109/LAWP.2020.2976515]

[16]   X. Li, T. Zhang, M. Wei, and L. Yang, "Reduction of truncation errors in planar Near-Field antenna measurements using improved Gerchberg–Papoulis algorithm", *IEEE Trans. Instrum. Meas.*, vol. 69, no. 9, pp. 5972-5974, 2020.
[http://dx.doi.org/10.1109/TIM.2020.3011588]

CHAPTER 4

Terahertz Technology Applied in Mobile Communications

Anthony J. Vickers[1] and **Jia Ran**[2,*]

[1] *Department of Electronic Systems Engineering, University of Essex, Wivenhoe Park, Colchester, Essex, UK*

[2] *College of Electronic Engineering, Chongqing University of Posts and Telecommunications, Chongqing, China*

Abstract: Owing to its ultra-high frequency, the terahertz band becomes one of the candidate bands for 6G communication. In this chapter, the characteristics and application fields of the terahertz wave, especially the application in wireless communication, are introduced in detail. Firstly, the characteristics of the terahertz wave are briefly introduced, and then the terahertz technology, including terahertz devices and the main application fields of the terahertz wave, is introduced. Next, focusing on the application of the terahertz wave in the wireless communication system, it introduces two mainstream terahertz wireless communication systems, describes the key technologies, and finally describes the potential application prospect of the terahertz wave in the wireless mobile communication system.

Keywords: Device, Direct modulation, Solid-state, Terahertz wave, Wireless communications.

INTRODUCTION

The development plan of 6G communications includes the technological path and societal path that stimulate new services. Near-future services in the sixth generation of mobile communication networks include holographic communications, high-precision manufacturing, artificial intelligence, the integration of subterahertz or visible light communication in a 3D coverage scenario. The technologies supporting above mentioned new services can be categorized into five parts, *i.e.*, a new internet architecture that combines kinds of resources within a single framework, a distributed AI algorithm, a 3D communication infrastructure, a new physical layer incorporating subterahertz bands and VLC, and a distributed security mechanism [1].

---

* **Corresponding author Jia Ran:** College of Electronic Engineering, Chongqing University of Posts and Telecommunications, Chongqing, China; Tel: 0086 23 62460592; Fax: 0086 23 62460804; E-mail: ranjia@cqupt.edu.cn

Xianzhong Xie, Bo Rong, Michel Kadoch (Eds.)

With the popularity of smartphones, the number of wireless network users and data demand is increasing rapidly. The rapid development of intelligent terminal applications requires that the future communication system can achieve ubiquitous ultra-high-speed access in a variety of complex environments. Therefore, one of the effective ways to improve spectrum efficiency is through advanced signal processing techniques and modulation schemes. However, due to the narrow bandwidth of the current operating frequency band, it is difficult to achieve a transmission rate of 100 Gbit/s. Another alternative is to use higher carrier frequencies to increase the channel bandwidth to provide sufficient transmission capacity.

Millimeter wave and terahertz band are the candidate frequency bands of high-frequency communication. They can cope with the problems faced by the current wireless communication system. In contrast, the terahertz band has greater potential than the millimeter wave band. Firstly, the bandwidth of the terahertz band is 0.1 ~ 10 THz, which is an order of magnitude higher than that of millimeter-wave, which can provide terabit per second data transmission rate support. Secondly, due to the decrease of antenna aperture, terahertz has higher directivity than millimeter-wave and is less prone to free space diffraction. Finally, the distance between transceivers in the terahertz band is much shorter than that in millimeter wave band, which will reduce power consumption and thus reduce carbon dioxide emissions [2]. Considering the shortcomings of the current communication system and the unique advantages of terahertz band, terahertz communication technology has attracted extensive attention in academic and industrial circles. It is considered to be the key wireless technology to meet the real-time traffic demand of mobile heterogeneous networks, which can alleviate the capacity bottleneck of the current wireless system and realize ultra-high-speed wireless communication. The huge bandwidth of the terahertz band and the super high-speed data transmission rate will realize a large number of new applications and services, such as vehicle communication, virtual reality (VR)/augmented reality (AR), health monitoring, satellite communication, *etc.*

According to the generation method of the terahertz wave, the current terahertz wireless communication equipment is divided into two parts. The first method is to use optoelectronic technology to convert optical frequency to terahertz frequency. That is, continuous or pulsed terahertz radiation is generated by semiconductor excitation. The second method is to use a frequency multiplier to increase the working frequency of electronic equipment from millimeter wave to terahertz range. The application of optoelectronic combination in terahertz wireless communication systems is often restricted by optical components, which is not conducive to the integration and miniaturization of chips. Therefore, the communication system based on frequency multiplier is widely used. However, there are still some difficulties in the large-scale application of terahertz communication systems, such as large volume and low integration of terahertz devices, high transmission loss of terahertz signal, and limited transmission power of terahertz RF devices. These problems require the industry to explore the development of new semiconductor materials and integrated circuit technology, research and development of advanced antenna technology, optimization of system resource allocation, and so on, so as to realize the miniaturization, low power consumption, and low cost of terahertz communication. The coverage of terahertz communication is enhanced, and the transmission rate of terahertz communication is improved. In order to better apply terahertz communication technology to support future ultra-high speed and low delay new applications, it is necessary to better capture the characteristics of terahertz band, understand the existing problems and technical challenges of terahertz communication, to build a more robust and efficient terahertz wireless communication system. This chapter firstly introduces terahertz technology, including terahertz wave and modulation devices, as well as its mainstream application scenarios. Secondly, the terahertz wireless communication technology is summarized, including the current terahertz channel propagation model, wireless communication system and mobile communication application scenarios. Finally, the possible important research directions of terahertz band in the future are prospected.

TERAHERTZ TECHNOLOGIES

Terahertz Wave

The term "terahertz" first appeared in the microwave field in the 1970s. It is used to describe the spectral frequency of interferometer and the frequency coverage of diode detector. At present, the electromagnetic wave with the spectrum of 0.1 ~ 10THz is named as terahertz wave. Its wavelength ranges from 30 μ m to 3000 μ m, which is between microwave and infrared light wave. It is located in the transition region between macro electronics and micro photonics.

With the development of mobile communication services, mobile users put forward higher requirements for wireless communication rate. According to Shannon's theorem, the channel capacity is proportional to its spectrum bandwidth. Therefore, larger bandwidth is the key factor to achieve ultra-hig--speed data communication. As a new band between microwave and light wave, terahertz has not been fully developed in the field of communication. Terahertz communication has the advantages of rich spectrum resources and high transmission rate, which is a very favorable broadband wireless access technology in future mobile communication. It can support higher data rates than millimeter band communication, ranging from tens of Gbps to several Tbps. Because terahertz wave is easily absorbed by water when it propagates in the air, it is suitable for high-speed wireless communication over short distances. It has the advantages of narrow beam width, good directivity, strong anti-jamming ability, and can realize secure communication within 2 ~ 5km. Terahertz wave can also be widely used in space communication, especially for broadband communication between satellite and ground. In outer space, there are transparent atmospheric windows near 350 μ m, 450 μ m, 620 μ m, 735 μ m and 870 μ m wavelengths of terahertz waves, which can realize lossless transmission and complete long-distance communication with minimum consumption. The wavelength of terahertz band is short, which is suitable for high-quality MIMO systems with more antenna arrays. The preliminary study shows that the beamforming and spatial multiplexing gain provided by high-capacity MIMO can overcome the rainfall attenuation and atmospheric fading of terahertz wave propagation, and can meet the coverage requirements of dense urban areas (such as 200m cell radius). Compared with wireless optical communication, terahertz wave has lower photon energy and can be used as carrier to obtain higher energy efficiency.

Compared with wireless optical communication, terahertz communication system is not sensitive to atmospheric effect in outdoor wireless communication. In indoor wireless communication, terahertz wave is easier to track beam than optical band, which can enhance the mobility of wireless communication system. In addition, terahertz communication system has an advantage, that is, by using reflection path, the link gain in indoor scene can be improved. Because of its unique characteristics, terahertz communication has many advantages over microwave and wireless optical communication, which determines that terahertz wave has broad application prospects in high-speed and short-range broadband wireless communication, broadband wireless security access, space communication and other fields.

Although terahertz wave has irreplaceable advantages in mobile communication, it still faces many challenges. The propagation characteristics of electromagnetic wave show that the amplitude of free space fading is proportional to the square of frequency, so the free space fading amplitude of terahertz is much larger than that of low frequency band. The propagation characteristics of terahertz wave and the large number of antenna arrays mean that terahertz communication depends on high directional propagation. According to the characteristics of the high directional signal, the corresponding mechanism needs to be designed and optimized. The attenuation of obstacles is another disadvantage of terahertz wave. For example, the attenuation of brick is as high as 40 ~ 80dB, and the human body can bring 20 ~ 35dB signal attenuation. However, the influence of humidity / rainfall fading on terahertz communication is relatively small. Under 100GHz, humidity / rainfall attenuation increases rapidly with the increase of frequency. In the future, several terahertz bands with small rain attenuation can be selected as typical terahertz communication bands.

Terahertz Devices

Terahertz Sources

Terahertz band is much higher than microwave band and lower than optical band. Therefore, it is difficult to generate terahertz wave only by using electronic devices that generate microwave signals or photonic devices that generate optical signals. Up to now, terahertz signals are usually generated by electronic devices through frequency doubling, or by photonic devices through optical mixing. There are two kinds of terahertz signal: pulse signal and continuous signal. Pulse terahertz signal is widely used in terahertz system, such as terahertz time domain spectrometer. There are usually five common terahertz sources. The first one is the quantum cascade laser, which uses advanced semiconductor heterostructure manufacturing methods to avoid the limitation of semiconductor band gap. Its semiconductor layer is very thin. When electrons tunnel from one layer to another, it will lead to very low energy transition. Terahertz wave is located in the emission radiation band. Since 2002, quantum cascade lasers have made rapid progress in frequency coverage, output power and operating temperature, which can produce terahertz wave sources of more than 10 mW. In recent years, CMOS has been developed rapidly. This technology has the advantages of high integration, small size and low cost. The high frequency capability of CMOS provides a solution to the problem of low frequency terahertz wave, which is realized by adding voltage controlled oscillator (VCO) or inserting active multiplication chain into CMOS devices [3]. The frequency of low frequency band can be doubled to terahertz band by using different frequency multiplier. CMOS terahertz circuit is realized by process scaling. In 2006, the power gain frequency of reference based on 65nm CMOS process has reached 420GHz. Uniaxial strained silicon transistors with physical gate length of 29nm have been put into use. The transistor cut-off frequency reached 485GHz with 45nm microprocessor technology in 2007. The chip consists of terahertz VCO and phase locked loop circuit, which enables the device to generate hundreds of gigahertz waveforms. A 40nm CMOS transmitter with a cutoff frequency of 300GHZ is used to achieve a data transmission rate of 105Gbps. Although HBT and HEMT technology can obtain higher high frequency source output power, CMOS is still an attractive candidate technology for terahertz technology due to its low cost and high integration density. The third commonly used terahertz source is the photoconductive antenna. The terahertz pulse generated by the photoconductive antenna has several unique characteristics, such as asymmetric intensity of positive and negative components of terahertz field, quasi half period characteristic, relatively low center frequency and low frequency band (usually 0.05-1THz). For the study of nonlinear terahertz phenomena, the low frequency of terahertz is conducive to the generation of strong fields, which can effectively drive electrons in many nonperturbative phenomena. Driven by this unique characteristic, coupled with the rapid development of Ti: sapphire amplifier lasers, people have begun to generate intense terahertz pulses from photoconductive antennas. A direct method is to increase the aperture of principal component analysis (PCA), so the generation of strong terahertz pulse by large aperture photoconductive antenna is studied. In fact, it is worth noting that the large aperture photoconductive antenna is the first method to generate strong terahertz pulse based on laser. Long before people pay attention to the strong terahertz source and its application, the optical ionization of high-power laser in the air can also generate strong terahertz wave. The advantage of this method is that the plasma in the air is a nonlinear medium, and the laser intensity is higher than the damage threshold of nonlinear crystal or photoconductive antenna. Hamster *et al.* demonstrated this technique for the first time. In these studies, femtosecond laser pulses with energy up to 50 mJ are used to ionize gas molecules. Later, DC bias is applied to the focused position of the laser beam by using electrodes to make the electric field perpendicular to the propagation direction of the laser. The DC electric field is used to accelerate the electrons in the plasma, resulting in the amplitude increase of coherent terahertz pulse. This technique is called DC bias. However, the air breakdown caused by high bias voltage limits the amplification of terahertz field. In the same period, the second harmonic bias method was proposed, which generated higher terahertz field based on the fundamental combination of pulsed laser and its second harmonic. Compared with monochromatic technique, the amplitude of terahertz field is increased by 40 times. The 800 nm femtosecond pulse generated by the amplified Ti: sapphire laser is focused into the gas to form plasma. A crystal that produces second harmonic, usually barium borate (BB), is placed in front of the focus and produces 400 nm of light. The generation process of terahertz is sensitive to the relative phase of the two colors, which can be adjusted by changing crystal position, using air dispersion or phase-shifting glass plate. Due to the excellent electro-optical properties of graphene, graphene-based technology has become a promising terahertz source technology. Metals are another promising material for terahertz sources because they exhibit wavelength independent large pump absorptivity, short electron lifetime (10 to 50 fs), non-characteristic terahertz refractive index (for seamless emission) and large thermal conductivity (effective removal of excess heat). In addition, the technology of metal film stack (heterostructure) is also very mature. The full potential of terahertz metal emitters is far from being realized.

Terahertz Modulators

Terahertz modulator is the key component of terahertz system, which has a wide range of applications from imaging to communication. The key to realizing high-speed communication with terahertz wave is to encode the information by using fast and effective amplitude and phase modulators in the carrier wave. In recent years, the discovery of metamaterials has provided new solutions for the flexible regulation of electromagnetic radiation in the terahertz band. The combination of metamaterials and semiconductor technology has made a significant breakthrough in dynamic terahertz functional devices, and has made remarkable achievements in terahertz amplitude and phase modulations. In addition, graphene and other new two-dimensional materials have laid a new foundation for the development of electronic and optical terahertz modulators.

Terahertz Amplitude Modulator

The amplitude modulation of terahertz wave is one of the most common modulation methods. For the design of terahertz amplitude modulator, there are two challenges: one is the modulation speed, the other is the modulation depth. Because natural materials cannot control terahertz wave effectively and rapidly, it is necessary to find suitable materials and structures to realize high-speed modulation of terahertz wave. At present, the materials used in terahertz modulator are 2DEG composite materials (such as GaN-HEMT) and two-dimensional materials (such as graphene). In terahertz band, artificial materials with response time less than 1 picosecond are the future development direction. In addition, the current mechanism is to combine the structure with semiconductors, resulting in high parasitic resistance and capacitance, which greatly limits the modulation speed. Therefore, the terahertz amplitude modulator should consider a new mechanism with more compact structure and lower parasitic parameters, and cannot completely rely on the mode concept of metamaterials or metasurfaces. In addition, the impedance matching circuit plays an increasing important role in the high-speed modulation above 10 Gbps, which affects the modulation speed and modulation speed. The basic modulation mechanism is one of the key points in the research of modulator. At the same time, more attention should be concentrated to the reasonable design of its control circuit.

Terahertz wave replication can be effectively controlled by injecting or exhausting carriers in semiconductor materials. In 2006, Amos lab in Las Vegas proposed an artificial micro structure Terahertz modulator. Based on the principle of Schottky diode, the modulation depth can reach 50% by changing the carrier concentration of split ring resonator. In addition to electrical modulation, similarly, the carriers generated by photons in semiconductors can be changed by optical doping of semiconductors. In 2010, N. h. Shen demonstrated a tunable blue shifted metamaterial optically realized in terahertz. The tuning range of the device can reach 26% (from 0.76 THz to 0.96 THz) by optical control of silicon [4].

A terahertz modulator combining high electron mobility transistor (HEMT) and SRR has been proposed for high-speed modulation performance. In 2011, D. Shrekenhamer firstly proposed an electronic controlled composite terahertz modulator based on HEMT / SRR, which can realize high-speed modulation of terahertz wave. For THz wave with frequency of 0.46 THz, the modulation speed of the device is about 10 MHz. Since then, various structures composed of metamaterials and HEMTs have been proposed. In 2015, Y. X. Zhang demonstrated a composite metamaterial structure based on an equivalent collective dipole array with a dual channel heterogeneity to obtain an efficient, ultra-fast, all electron gated terahertz modulator. For the first time, the terahertz modulator realizes the real-time dynamic test of 1 GHz modulation speed and 85% modulation depth.

In recent years, many novel photoelectric response materials, such as graphene, have been used in terahertz modulators. In 2014, researchers at Rice University employed single-layer graphene with metal ring holes and applied grid voltage to achieve 50% modulation depth. In 2015, researchers from Nanyang University of Technology reported a monolithic integrated device of graphene and terahertz quantum cascade lasers, which achieved a modulation depth range of 94% - 100%, and a modulation rate of 100MHz in a specific region. In 2017, researchers from the University of Maryland in the United States covered the surface of passive silicon dielectric waveguide with graphene film, and the maximum modulation depth was more than 90% under the external gate voltage. Vanadium dioxide (VO_2) with phase transition characteristics is also widely used in terahertz modulators. For example, researchers from the University of Electronic Science and technology of China have successfully developed a terahertz modulator with a modulation rate of 10 MHz and a modulation depth of more than 85%. Researchers at the University of Pennsylvania have prepared a digital memory that uses the phase transition of vanadium dioxide to eliminate read and write records. In addition, other new semiconductor materials are also used in the design of terahertz modulator. In 2016, researchers from Nankai University realized an optically pumped terahertz modulator using a new two-dimensional material MoS_2 [4].

Terahertz Phase Modulator

In addition to amplitude modulation, terahertz phase modulator is also one of the research focuses of terahertz modulator. The first terahertz phase modulator based on undoped GaAs metal split cavity metamaterial structure was successfully developed in 2009. A 30 degree phase shift is achieved by electrically tuning the carrier density near the SRRs gap and changing the L-C resonant mode. Later, in a GaAs loaded structure, an external ultrashort laser pulse was used to change the dipole like resonance in the structure to achieve a phase shift of 40-50 degrees. In addition, a switch controlled asymmetric double split ring structure with 32.2 degree phase shift is proposed.

Although these electrically tunable and optically tunable materials can achieve certain phase modulation, it is still a severe challenge to obtain larger phase shift. The application of new materials brings new opportunities to solve this problem. In 2016, Y. Urade proposed a reconfigurable planar metamaterial, which can switch between the capacitive response and the inductive response through the local change of the conductivity of the tunable material. Based on Babinet principle, the device utilizes the singular electromagnetic response of metal checkerboard structure. The Babinet inverter metasurface with thermostat can be used as switchable filter and switchable phase shifter. Like amplitude modulator, HEMT is also suitable for THz phase modulator design. In 2017, HEMT (high mobility transistor) was successfully developed. At 4V reverse gate voltage, the terahertz modulator has 80% modulation depth at 0.86 THz and 38.4° phase shift at 0.77 THz. At the same time, the high switching speed of HEMT means that terahertz modulator has high modulation speed potential. In 2018, in order to achieve large phase shift, a VO_2 loaded dumbbell composite resonator was proposed. The results show that there is a mixed mode with enhanced resonance intensity, which is formed by the coupling of L-C resonance and dipole resonance. By using the photoinduced phase transition of VO_2, the resonance intensity of the mode is dynamically controlled, which makes the incident terahertz wave produce a large phase shift. The dynamic experimental results of China Telecom in February 2019 show that the phase shift of 138 degrees can be achieved by controlling the power of external laser through this single-layer VO_2 loaded composite structure. In addition, the phase shift exceeds 130 degrees in 55GHz (575-630 GHz) bandwidth.

The discovery of metamaterials has greatly improved our ability to regulate terahertz electromagnetic radiation. The combination of metamaterials and semiconductor technology has made a great breakthrough in dynamic terahertz modulator. Metamaterials and metasurfaces achieve new electromagnetic properties and functions by cutting subwavelength structures and integrating functional materials. For phase modulators, in addition to modulation speed, how to achieve a large linear phase shift is another great challenge.

Terahertz Antennas

With the help of the newly invented terahertz light source, terahertz system has penetrated into various applications. One of the most important technologies in terahertz communication system is to design an efficient terahertz transceiver antenna. In the current terahertz system, antenna is the key component of transmitting and receiving terahertz wave. In traditional mode, the signal source and antenna are usually regarded as two independent parts. While in terahertz system, the source and radiation parts may need to be considered as a whole, especially for terahertz time-domain spectroscopy system.

Antenna design for terahertz band sources requires more consideration than low frequency antenna design. Photoconductive antenna is the most widely used antenna in terahertz time domain spectroscopy system. This method was firstly proposed by Auston and his colleagues in the early 1980s. The schematic diagram of terahertz time domain spectrum system is shown in Fig. (1). The 800 nm laser is directed to the beam splitter, which divides the laser into pump pulse and detection pulse. Terahertz radiation is generated on the photoconductive antenna by pumping pulse, and then refocused by off-axis paraboloid mirror. After passing through the sample, the terahertz wave is focused on the detector again. The probe pulse is delayed by a delay line to generate a delay compared to the pump pulse. The duration of laser pulse is femtosecond, which is much narrower than that of terahertz picosecond. This detection mechanism can provide enough time-resolved sampling rate.

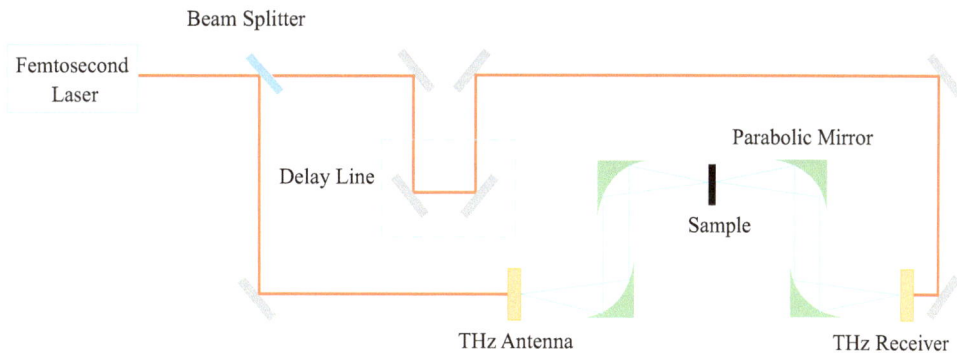

Fig. (1). Schematic diagram of terahertz time domain spectroscopy system.

The horn antenna is also one of the most commonly used antenna types in terahertz band. In Atacama large/submillimeter array (ALMA) and Planck telescope, horn antennas are used as feeds. They are arranged on a focal plane.

More than 12 feeding modes working at different frequencies. The frequency range of Planck array horn is 27.5-870 GHz, which is 30 GHz, 44 GHz, 70 GHz, 100 GHz, 143 GHz, 217 GHz, 353 GHz, 545 GHz and 857 GHz respectively. In fact, terahertz horns are usually connected to waveguides, which previously specified a maximum frequency of 330 GHz. The newly approved standard rectangular metal waveguide for terahertz applications has extended the operating frequency to 5 THz. In addition, the physical size of terahertz waveguides is in the micron range, which brings challenges to the manufacturing industry [4].

Reflector antennas are widely used from microwave communications to optical telescopes. Reflector antennas are usually electrically large and have diameters ranging from several wavelengths to hundreds of wavelengths. This electronic size usually provides a high gain or high efficiency receiving area. In general, electrically large antennas provide very narrow beams, 3dB beam width is usually several degrees or even narrower. This kind of narrow beam has high gain because of its energy concentration, which is of great significance in remote sensing applications. Although the single reflector antenna is widely used in microwave and millimeter wave, it is rarely used in terahertz band. One example is the installation of the main antenna on the payload of advanced microwave detection unit B (AMSU-B). In terahertz band, the parabolic reflector is Cassegrain type or Gregory type. Planck telescope adopts bias structure and Herschel telescope adopts feed-forward structure. The Planck telescope's main antenna is 1.5 meters in diameter, while the Herschel telescope's main disk is much larger with a diameter of 3.5 meters. The antenna can detect cosmic background radiation and deep space signal with a gain of 70dB. Ground based observatories also use electrically large reflector antennas or arrays. The ALAMA antenna array has an antenna with a diameter of 7 meters or 12 meters. Such an array forms an interferometer that can detect signals from deep space with an angular resolution of 0.1 acre. The frequency range of ALAMA array is 31-950GHz. Spherical centers are also used, such as FAST primary mirror under construction in China and Arecibo Observatory in Chile [4].

The potential application of terahertz wave is sensor and communication. In order to realize high-speed tracking and aiming, dynamic beam control is also needed in terahertz band, and dynamic scanning antenna is also an urgent problem to be solved. Early attempts have been made to scan beams at very high frequencies, including the ongoing study of reflective arrays based on 120 GHz MEMS. These frequencies can be considered to be the upper limits of the applicability of standard RF-MEMS technology, as MEMS components become too large to be effectively integrated into array element elements. Therefore, it is very important to study the technology based on reconfigurable materials. Common reconfigurable materials include liquid crystal (LCS) and graphene. Liquid crystal is a kind of common reconfigurable material, and its anisotropic permittivity tensor can be modified by the applied bias field. In the laboratory, the reflection coefficient of liquid crystal can be regulated by bias voltage using the reflection array loaded with LCS. In addition, terahertz beam scanning can also be realized by using graphene. In this case, the dynamic phase control is realized by applying bias voltage on the nearby electrode (so-called graphene field effect). These concepts of LC and graphene structure are based on the typical resonant unit topology. For LC, the substrate parameters are controllable, while for graphene, the substrate is fixed, but the resonance can be changed by changing the composite conductivity of graphene sheet.

As the most common transceiver antenna in terahertz system, the main drawback of photoconductive antenna is low power conversion efficiency. Although the photoconductive antenna is called antenna, it is also an integrated terahertz signal generator. In order to improve its directivity and radiation power, various methods have been explored. At present, there are some solutions to improve the performance of photoconductive antenna: 1) using laser pump source with high intensity and narrow pulse width; 2) Optimizing the semiconductor substrate material; 3) changing the antenna structure to increase the breakdown voltage of the substrate; 4) improving the transmission efficiency of terahertz wave on the antenna; 5) improving the spatial coupling efficiency of terahertz wave; 6) using anti-reflective layers such as silicon nitride, silicon dioxide, indium tin oxide on the surface of the semiconductor substrate to reduce the reflection of the incident light power, but this method is relatively expensive; 7) integrating metamaterials into photoconductive antennas, This method has two advantages: one is to expand the optimization freedom of photoconductive antennas by introducing metamaterials. The other is to optimize the design of metamaterials for one performance without affecting other parameters. Other high gain antennas, such as horn antenna and reflector antenna, often have fabrication problems in terahertz range due to the short wavelength of high frequency. It is eager to find a proper method to design terahertz antennas with high efficiency, low cost and high gain.

Terahertz Detectors

The photoconductive antenna is also a coherent terahertz wave detector. The detection mechanism of terahertz wave by photoconductive antenna is similar to that of photoconductive antenna, which is an inverse process without bias voltage. In the principal component analysis (PCA) detector, femtosecond laser pulses illuminate the gap between the principal components and excite the carriers to the conductive band. Similar to terahertz generation method, terahertz wave can be detected by laser-induced air plasma technology. The strong terahertz pulse may also be ultra-wideband (bandwidth 0.1-200 THz), so it is necessary to measure and cover the whole spectrum. One of the challenges in measuring ultra-wideband terahertz pulsed electric fields is the need for high-power lasers, which usually have limited repetition rates. Compared with the high repetition rate laser, the high-power laser has larger beam fluctuation, which is also reflected in the inter beam stability of strong terahertz pulse, especially when using nonlinear optical technology. Therefore, the traditional scanning probe technology is difficult to measure terahertz field, which usually needs several minutes to several hours of stable terahertz output, which depends on the terahertz field, laser repetition rate and scanning step size. In addition, single pulse UWB measurement technology is needed to accurately characterize these strong terahertz pulses in many cases. This single terahertz measurement technique will also be very useful for some experiments involving irreversible processes. Terahertz absorber is another kind of terahertz detector. Though it lacks conversion from light to electrical signal, it can absorb all 100% terahertz wave. As Fig. (**2**) shows, metasurface loaded with graphene is a typical terahertz absorber to realize high and ultra-wideband absorption. Besides, traditional Nipkow disks can also be used as terahertz detector for imaging [5].

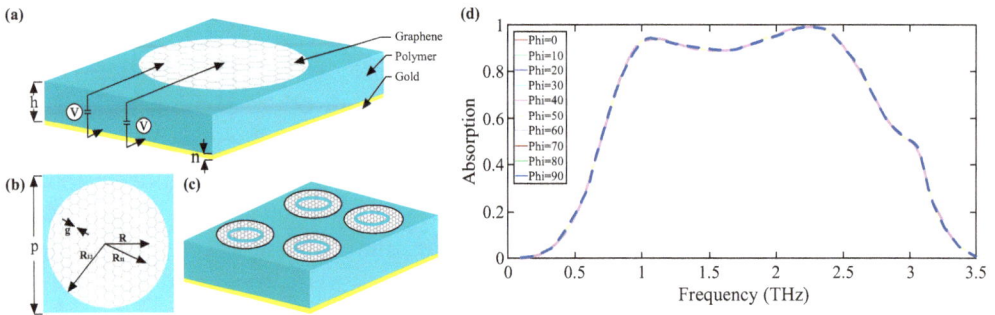

Fig. (2). Schematics of the terahertz absorber (a-c) and its absorption depending on polarization angle.

Applications

Imaging

As an electromagnetic wave, terahertz light wave has imaging characteristics similar to visible light and infrared light. It can obtain good imaging results in coherent and incoherent fields. In particular, terahertz wave has specific infrared band characteristics, such as penetrability and hydrophilicity, to detect hidden targets.

According to different imaging methods, terahertz imaging can be divided into passive imaging and active imaging. Passive imaging mainly detects terahertz radiation of the sample itself. There is a positive correlation between the intensity of terahertz radiation and temperature. The human body temperature is usually higher than the ambient temperature, so the human body will radiate stronger terahertz wave, which is very different from the radiation characteristics of metal, dielectric and other materials. Due to the strong penetration of terahertz wave, ordinary cloth, paper, *etc.*, can be detected quickly. Although the principle of passive imaging is relatively simple, there are still many problems in practical application. Firstly, due to the influence of the sensitivity of existing terahertz detectors and photon noise, the terahertz radiation emitted by human body is generally weak and the imaging resolution is low. Secondly, due to the interference of environment to terahertz radiation, the imaging noise of the system is generally large and the signal-to-noise ratio is low. In order to improve the reliability and practicability of passive terahertz imaging system, researchers have developed a variety of image enhancement methods for passive terahertz imaging system, such as wavelet denoising, histogram equalization, edge filtering, image segmentation, *etc.* These methods can improve the image quality and reduce the image noise to a certain extent. For active terahertz imaging system, the imaging quality is greatly improved compared with passive terahertz imaging due to its high radiation power. At the same time, the strong coherence of terahertz radiation beam makes it develop rapidly in the fields of holographic imaging, CT imaging, lensless imaging and terahertz interference imaging. Active terahertz imaging system usually includes: stable continuous terahertz generator, beam shaping module, focusing optical path, terahertz detection optical path (usually using focal plane array FPA) and mechanical scanning control system. According to different imaging methods, active terahertz imaging can be divided into transmission imaging and reflection (scattering) imaging. Compared with the transmission system, the reflection system can not only recognize the hidden target, but also obtain the surface features of the hidden target. Especially in the small security inspection equipment, the reflection system can greatly reduce the volume of the security inspection equipment, and effectively reduce the development cost of the system.

Spectral Analysis

Because the terahertz light wave can have strong resonance interaction with the vibration energy levels and chemical bonds of liquid and solid macromolecules, the spectral analysis of some dangerous goods, such as explosives, drugs and knives, can be realized by using terahertz wave. Many biological macromolecules, such as protein molecules and DNA bases, whose characteristic vibration frequency is just in the terahertz frequency band, show strong resonance absorption to terahertz wave too. That is to say, terahertz wave can characterize the fingerprint spectrum of substances. Meanwhile, terahertz radiation will not ionize the matter, proving that it is a good candidate for nondestructive testing. These excellent characteristics make them very suitable for the sensing and measurement in the field of biochemistry, manufacturing industry and so on.

Communications

Terahertz band can offer ultra-high data rate. Due to the limitation of transmission distance, terahertz wireless communication can be used in indoor ultra-high data-rate application scenarios, including WPAN, WLAN and indoor cellular network. Generally, WPAN system is expected to achieve specific applications such as mobile phones, laptops, headphones and other tablet devices. The current low data speed wireless communication seriously limits the development of virtual reality technology. Once terahertz bandwidth is used in wireless communication system, VR technology will bring better user experience than wired communication system. In addition, terahertz band communication is also used in indoor cellular networks with large-scale cellular structures. In indoor cellular networks, both static users and mobile users can enjoy ultra-high speed data communication services in such a small cell. However, due to the path loss, maintaining the reliability of terahertz wireless link is a problem which is worth exploring. In addition to the possible indoor terahertz application scenarios, secure wireless communication is another important application in terahertz wireless communication. Owing to its very narrow beam and high directional property, eavesdropper that is not in the propagation path can not wiretap the secure intelligence. Meanwhile, its massive bandwidth makes it be hidden to jammers. Spread spectrum technology and frequency hopping technology can also be used to counter jamming attacks.

Difference with Other Communications

Terahertz wave length is between millimeter wave and infrared wave, which has its specific communication advantages. Compared with millimeter wave, terahertz wave can obtain higher data transmission rate and link directivity, terahertz waves have shorter wavelengths than millimeter waves, so free space diffraction is smaller. Therefore, the use of small antenna with good directivity in terahertz communication can reduce the transmission power requirements and reduce the signal interference between different antennas to a certain extent. Another interesting feature is that in terahertz band, the probability of eavesdropping is lower than that of millimeter wave. This is due to the high directivity of terahertz beams, which means that unauthorized users must intercept messages on the same narrow beam width. In wireless communication, using infrared radiation is one of the most attractive and promising choices for radio spectrum. In inconvenient weather environments such as fog, dust and turbulence, terahertz band is a good alternative to infrared communication. Compared with infrared band, fog attenuation in terahertz band is smaller. Recent experimental results show that atmospheric turbulence has a great influence on infrared signal, but has little effect on terahertz signal. In addition, the attenuation of cloud dust makes the infrared channel smaller, but has little effect on terahertz signal. In terms of noise, terahertz system is not affected by ambient light signal source. Due to the low photon energy in terahertz band, thermal noise also contributes a lot to the total noise.

Visible light communication (VLC) has attracted much attention due to its advantages in power consumption, volume, cost and duration. It can support indoor positioning, human-computer interaction, communication between equipment, vehicle network, traffic lights, advertising display and other important business and applications. The data transmission rate largely determines the consistency of the field of view between the receiver and the receiver, but this is difficult to maintain due to the influence of the receiver's motion and direction. In addition, the interference of ambient light can significantly reduce the received signal-to-noise ratio and the communication quality. Terahertz band is considered as a candidate frequency band for uplink communication, which is a capability lacking in VLC communication. In addition, the deep ultraviolet band (200-280nm) has been proved to be a natural candidate for short-range NLOS communication (also known as light scattering communication). Compared with the deep ultraviolet link, terahertz band is considered to be the biggest competitor. Different from UV communication, UV communication limits the health and safety of eyes and skin, while terahertz band is non ionizing, which is very friendly to users' health.

Terahertz wave communication has many advantages, such as high transmission rate, high anti-interference, environment-friendly, *etc*. It combines the advantages of millimeter wave, infrared and visible light communication, and is one of the powerful communication candidates.

TERAHERTZ WIRELESS COMMUNICATIONS

Terahertz Wireless Communication Systems

Although the THz wireless communication system has great advantages compared with traditional communication system, THz wireless communication system was not developed until the 21st century due to the lack of necessary technical support conditions. Solid state terahertz communication system and space direct modulation terahertz communication system are two commonly used terahertz communication systems. The former is based on mixing mechanism, while the latter directly modulates baseband signal into a radio frequency signal.

Solid State System

Although terahertz wireless communication has the advantages of high transmission rate, anti-interference and low noise, its development has been slow due to the lack of necessary hardware support. In recent years, wireless communication systems are mainly based on optoelectronic combination technology. However, this method is limited by the performance of optical components, which is not conducive to the integration and miniaturization of the system. The integration and preparation of terahertz communication system will be a serious challenge. Several typical all solid state terahertz communication systems are described below.

One of the most typical systems of all solid state terahertz communication is the solid state terahertz wireless communication system developed by the Nippon Telegraph and Telephone Public Corporation (NTT) in 2008 [4]. The maximum communication distance is 3-4km. In the next few years, NTT has been upgrading the system and replacing all optical excitation devices with InP HEMT MMICs, the maximum data transmission rate of 11.1Gbps can be obtained, and the error free transmission of 10Gbps signal above 800m has been realized. The system can also realize 10 Gbps bidirectional data transmission rate and 20 Gbps one-way data transmission rate. In 2013, Karlsruhe Institute of Technology (Kit) implemented a 0.24 terahertz all solid-state wireless communication system using integrated hardware circuit composed of high electron mobility transistors. The maximum transmission rate of the system reached 40 Gbps and the transmission distance exceeded 1000 meters. In 2016, University of Electronic Science and technology in China developed a 220GHz all solid-state wireless communication system experimental system, which realized real-time 3D HD video signal transmission of more than 200m and 3.52Gbps in outdoor environment. When the data transmission rate was 3.52Gbps, the bit error rate was. Cassegrain antenna is used in the receiver and transmitter of the system, and 220GHz subharmonic mixer is used for modulation and demodulation.

Direct Modulation System

All solid-state THz communication system can adopt QPSK or 16 quadrature amplitude modulation (16-QAM) modulation mode to realize ultra-high data rate transmission, which is the most popular wireless communication mode at present. In recent years, with the rapid development of high-speed terahertz modulators, direct modulation terahertz communication system has attracted more and more attention. Among them, direct modulation of THz communication system becomes an effective way to realize long-distance data transmission.

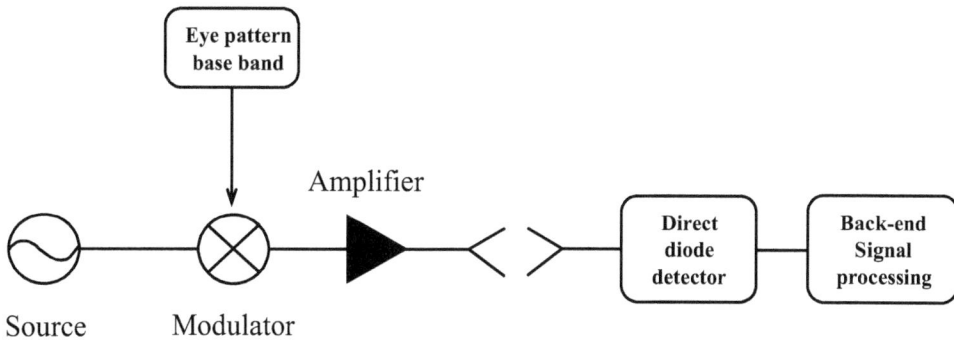

Fig. (3). schematic diagram of direct modulation terahertz system.

As shown in Fig. (**3**), in the first step, a continuous terahertz carrier wave is generated from a terahertz transmitter, which can be a solid-state electronic source or a vacuum electronic source. Then, the ASK modulation signal of serial digital HD video signal from baseband is loaded on the modulator to realize the on-off keying (OOK) modulation of THz carrier. At the same time, the THz wave is transmitted vertically to the modulator, resulting in OOK modulation signal loaded on the carrier. In the third step, the modulated THz wave is radiated by a parabolic antenna in space and received by another antenna. Because the terahertz wave modulated by OOK can be sampled directly, the system can use either terahertz detector or terahertz mixer as receiver. Finally, the received signal is demodulated and digitally decoded.

The first laboratory level THz communication system is developed by using the self-developed source (multiplier chain), modulator and detector. The system allows the out of space modulator to work at high power and has significant advantages in long-distance wireless data transmission. Therefore, direct modulation terahertz system can be applied to all kinds of terahertz sources, including continuous wave (CW) sources. In addition, the system directly loads the data to THz carrier through modulator, thus eliminating the adverse effect of phase noise deterioration when harmonic mixer is used as data loading converter. Therefore, without the use of terahertz amplifier, the system can output 5-10 MW modulated terahertz signal, which is two orders of magnitude higher than that of mixing and photonics mechanism systems.

The performance of terahertz communication system largely relies on terahertz modulator. However, although the performance of terahertz modulator is developing rapidly, the modulation speed is still lower than 3 Gbps, which limits the application of this direct modulation terahertz communication system. In addition, the device linearity of the modulator is not very good, which can not realize high-order modulation, which greatly reduces the bandwidth availability ratio of communication system. In short, the realization of high-performance terahertz modulator is a key point to develop a direct modulation terahertz communication system [4].

Key technologies in Terahertz Communications

Transceiver

According to the generation mode of terahertz signal, terahertz transmitter can be divided into pure electronic device transmitter, optical heterodyne transmitter and semiconductor laser transmitter.

Pure Electronic Devices

This kind of terahertz transmitter consists of radio frequency source (RF source), multiplier, mixer and amplifier. RF signal source generates RF signal with frequency of about tens of GHz, and passes through frequency multiplier. Electrical signal mixer is applied to generate modulation signal. The signal is amplified and sent by antenna, as shown in Fig. (**4**). The advantage of this method is that the transmitter structure is very simple, but the requirements for the bandwidth and gain of electronic equipment are restrict. There is a large conversion loss in modulation and demodulation, and the equipment is usually more expensive.

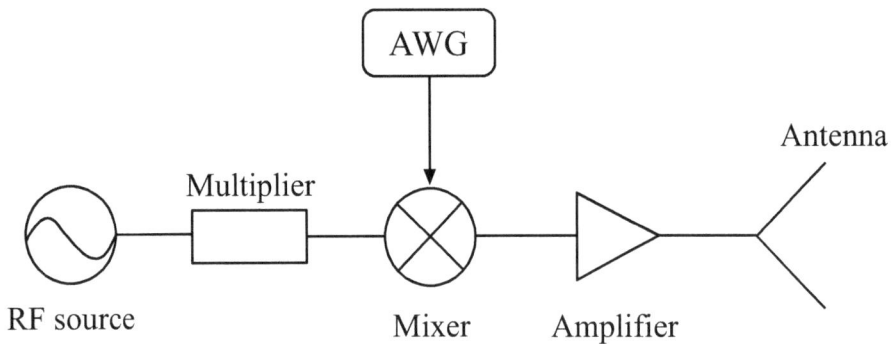

Fig. (4). Pure electronic transmitter.

In 2008, Braunschweig Laboratory of Germany successfully realized the real-time transmission of analog video signal on 300 GHz channel. In 2011, an all solid-state THz communication system working at 220 GHz is developed. The transmitter and receiver front-end uses active multi-functional millimeter wave integrated circuits (MMICs) to integrate frequency multiplier, mixer and amplifier, which can realize 256 QAM quadrature amplitude modulation (QAM) signal transmission, on off keying (OOK) signal rate can reach 25 Gbit / s, which is a successful attempt to use pure electronic mode for THz communication. Real-time wireless transmission of 1.485 Gbit/s high definition TV (HDTV) signal on 240 GHz and 300 GHz carrier frequencies is achieved. Bell Laboratories in the United States has realized a 625 GHz THz communication system using all electronics, which is a new breakthrough in carrier frequency. The system uses dual binary baseband modulation, which can achieve a transmission rate of 2.5 Gbit / s, and realize error free transmission. In 2012, karlsruhe Institute of technology used MMIC to achieve 25 Gbit/s 10 m wireless OOK signal transmission on 220 GHz carrier frequency.

In China, in 2011, China Academy of Engineering Physics realized 10 Gbit / s 16 QAM signal wireless communication in the 0.14 THz frequency band, at the same time, real-time transmission and demodulation of 4-channel HDTV signals were carried out, and the transmission distance reached 500 m. In 2014, China Academy of Engineering Physics realized the first 0.34 THz data link under 50 m line of sight channel using pure electronic devices, which can modulate and demodulate 3 Gbit / s data in real time. Based on this link, the researchers also proposed a 0.34 THz wireless local area network (WLAN) communication protocol based on IEEE 802.11b/g protocol, which is the preliminary verification of the feasibility of THz wave used in WLAN. China Academy of Engineering Physics has implemented a 0.14 THz OOK communication system with a transmission rate of 15 Gbit/s. The modulation and demodulation of the system can be directly completed by analog devices, and combined with high-order QAM modulation, harmonic mixing and cascade amplification technology. Compared with OOK modulation, it has high spectrum efficiency and can be used for channel estimation and equalization using advanced digital signal processing technology.

At present, the carrier frequency generated by frequency doubling method is 625 GHz, which requires up-down conversion and a variety of modulation mixing technologies. The conversion loss of electronic devices makes the transmitting and receiving power lower, so the transmission rate is not high and the wireless transmission distance is relatively close. OOK modulation is mainly used in modulation format, which belongs to lower order modulation and has low spectral efficiency. Some researchers have tried orthogonal modulation such as QPSK and QAM modulation.

Optical Heterodyne Method

The second scheme to generate terahertz wave is photon assisted method, namely optical heterodyne method. At present, most terahertz communication systems use this method, and the frequency of terahertz wave is below 1THz. The transmitter consists of an external cavity laser (ECL), an optical modulator, a polarization maintaining optical coupler (PM-OC), a polarization maintaining erbium-doped fiber amplifier (PM-EDFA), a variable optical attenuator (VOA), and an electrical amplifier (EA). It consists of amplifier (EA) and uni-traveling carrier photo detector (UTC-PD). The laser generates two or more optical signals, modulates the information to the optical carrier, and then combines with UTC-PD to generate THz signal with the frequency difference of two beams by optical heterodyne method,

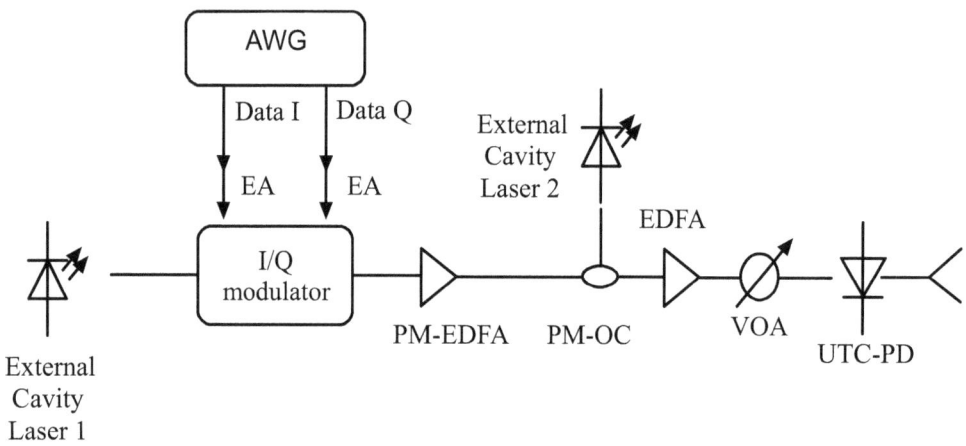

Fig. (5). Optical heterodyne transmitter.

As shown in Fig. (5), if an optical signal (optical frequency comb) with a certain frequency interval can be generated, then the frequency difference of any two light beams can be selected as the frequency of terahertz signal. The advantage of this method is that the frequency can be adjusted easily and terahertz signals of different frequencies can be obtained conveniently. The system capacity can be improved by using this method in multi carrier system.

In 2013, Koenig S. *et al.* made transmission rate reach 100 Gbit/s and wireless transmission distance of more than 20 m in 273.5 GHz frequency band. In 2014, Shams H. *et al.* realized multi input multi output (MIMO) system by combining polarization multiplexing technology, and realized real-time transmission of 75 Gbit/s dual channel QPSK signal in 200 GHz frequency band. In 2014, the University of Lille used THz optical mixer to realize 46 Gbit/s wireless transmission at 400 GHz carrier frequency, and the transmission power was less than 1 μ W. In 2015, Shams H. *et al.* implemented a four carrier downlink transmission system using an optical frequency comb with adjustable gain. The overall downlink channel data rate reached 100 Gbit/s, indicating that multi carrier transmission can increase the data rate and reduce the bandwidth requirements of photoelectric conversion devices. Denmark University of technology has implemented a 400 GHz terahertz transmission system, which can monitor and transmit 60 Gbit/s wavelength division multiplexing (WDM) QPSK signal in Nyquist channel in real time, which is the highest rate that can be realized on carrier frequency higher than 300 GHz at that time. The coherent radio over fiber (CROF) THz communication link has been implemented by Essen University in Duisburg, Germany, which can support offline 59 Gbit/s 64 QAM orthogonal frequency division multiplexing (OFDM) signal transmission; the link can also transmit HDTV signal in real time at 328 GHz carrier frequency. In the traditional laboratory environment, the rate of more than 100 Gbit/S is usually realized by space division or frequency division multiplexing technology, which increases the complexity and energy consumption of the system. In 2017, the Danish University of technology implemented a single channel 0.4 THz optical wireless combined terahertz link, using a group of transmitters and receivers to achieve a wireless transmission rate of 106 Gbit/s, with a wireless transmission distance of 50 cm.

Semiconductor Laser

The THz signal generated by pure electronic device or photon assisted method is usually at hundreds of GHz. In higher frequency band (such as 1 THz or above), laser method can only be used, such as quantum cascade lasers (QCL), gas laser and free electron laser to generate THz signal. QCL implemented by semiconductor technology is easy to integrate, which has aroused extensive research interest. Terahertz quantum cascade laser has the advantages of small volume, compact structure and short carrier lifetime, which can directly modulate the laser at high speed. The generation scheme is shown in Fig. (6).

In 2007, Barbieri *et al.* adjusted the width of the spectrum sideband of THz wave by directly modulating the bias voltage of the 2.8 THz QCL bimetallic waveguide, and the maximum spectrum sideband of 13 GHz could be generated. This is an early attempt of terahertz QCL. In China, the Chinese Academy of Sciences has paid most attention on terahertz QCL. In 2008, Shanghai microsystems Institute of Chinese Academy of Sciences realized wireless communication experiment based on terahertz QCL and terahertz quantum well photodetectors (QWPs) at 4.1THz frequency, transmitting pictures and sound signals, and further transmitting video signals in 2012. The advantage of THz QWPs lies in its short carrier lifetime and short transient response time. In 2013, the Institute of microsystems and information technology of Chinese Academy of Sciences has carried out audio, file and video communication experiments based on QCL and QWPs in 4.1 THz band. The system adopts OOK digital modulation format with transmission distance of 2.2 m. Ultra high frequency THz signal can be generated by semiconductor laser. However, the performance requirements of THz emission source are very strict. The source often needs to work at ultra-low temperature, which limits the availability of this method.

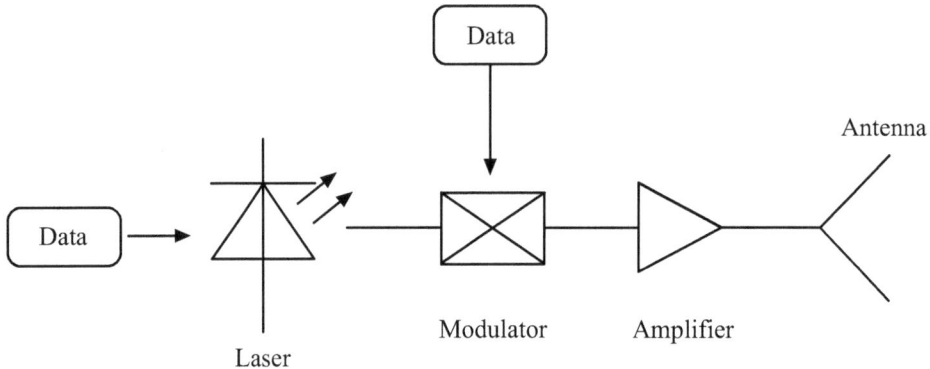

Fig. (6). Laser based transmitter.

Detector

The current terahertz receiver can be divided into direct detection terahertz receiver and superheterodyne detection receiver.

Direct Detection Terahertz Receiver

Fig. (7) shows a typical direct detection method using commercial grade Schottky barrier diodes with a cutoff frequency of 1-10THz. Direct detection method is often used in direct modulation terahertz wireless communication system. Direct modulation terahertz communication system is an effective way to realize long-distance data transmission. The system can output 5-10mW modulated terahertz signal, which is two orders of magnitude higher than that of mixing and photonics mechanism systems. Direct modulation THz communication system can achieve 1Gbps data transmission. However, the performance of terahertz communication system largely depends on terahertz modulator. Although the performance of terahertz modulator is developing rapidly, the modulation speed is still lower than 3Gbps, which limits the application of this direct modulation terahertz communication system. In addition, the device linearity of the modulator is not very good, which can not realize high-order modulation, which greatly reduces the bandwidth availability ratio of communication system. In short, to develop a direct modulation terahertz communication system, the realization of high-performance terahertz modulator is a crucial issue.

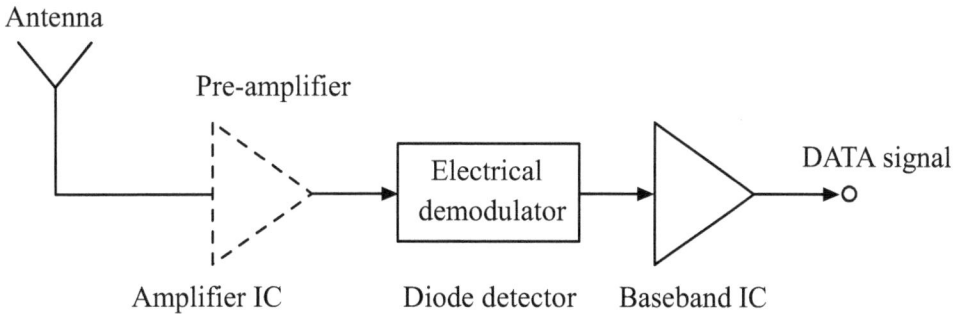

Fig. (7). Direct detection receiver.

Superheterodyne Detection Receiver

Compared with direct detection, superheterodyne receiver is more widely used. As shown in Fig. (8), superheterodyne detection using Schottky barrier diode mixer and LO signal source can provide high sensitivity. The preamplifier with low noise figure improves the sensitivity of the receiver and is verified in the frequency range of 300-400GHz. The core circuit of superheterodyne receiver usually includes frequency conversion circuit, signal generation circuit and amplification circuit. However, due to the immaturity of III-V compound semiconductor transistor technology, solid-state amplifiers are relatively scarce in terahertz band. The amplifier is mainly divided into power amplifier and low noise amplifier.

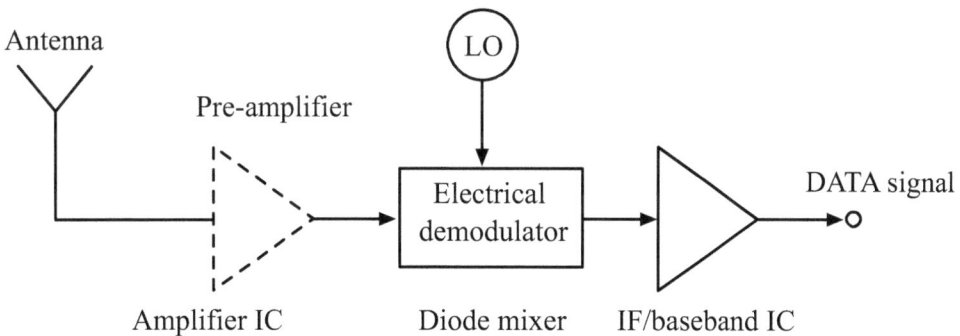

Fig. (8). Superheterodyne detection receiver.

As the last stage of the transmitter, the power amplifier is directly connected to the antenna. The size of the power amplifier directly affects the output power of the transmitter, and also directly determines the working distance of the system. Although LNA is usually used as the first stage of the receiver, it will affect the sensitivity and noise performance of the receiver. Only a few countries have the ability to implement this program, solid-state terahertz amplifiers develop at high prices and low production rates. Therefore, in the field of solid-state terahertz technology, scholars pay more attention to the research of mixer and multiplier based on Schottky barrier devices. The development of mixer and multiplier based on Schottky barrier diode is essentially the development history of Schottky barrier diode. Due to the lack of terahertz amplifier, mixer becomes the first stage of receiver, which affects the system performance. In terahertz band, subharmonic mixer is often used because it can reduce the difficulty of LO. The development of THz subharmonic mixer has experienced three stages of diode development, from the early use of whisker contact diode to single planar diode, and now integrated planar diode. Similarly, the development of multiplier is accompanied by different stages of diode development. Terahertz frequency multipliers usually use double or triple frequency. High order frequency multipliers are difficult to use because of their low efficiency. Frequency multiplier is more common than frequency multiplier in terahertz band, because the frequency multiplier needs to set a special loop for the second harmonic in the circuit, which increases the complexity of matching network of frequency multiplier. The realization of terahertz system depends on the breakthrough of key solid state circuits. In other words, the development of solid-state terahertz communication system requires in-depth research and exploration of mixers and multipliers.

Beamforming and Beam Tracking

Affected by spreading loss and molecular absorption loss, the transmission loss of terahertz wave is very large, and the transmission distance is seriously limited. The loss can reach 80dB for transmission distance of 1m, which greatly limits the promotion of terahertz wireless communication. Improving the gain and directivity of the antenna can effectively extend the transmission distance.

Using phased array antenna is one of the effective methods. By precoding, the radiation phase and intensity of the antenna array element are adjusted to obtain high directivity and high gain beamforming. Thus, the transmission distance of terahertz wave is extended.

Precoding mode, traditional beamforming methods in mmWave range include digital, analog and hybrid methods. Digital beamforming needs to provide dedicated baseband and and RF hardware for each antenna unit. The cost is very high, and the loss of mixed signal is also very large. Analog beamforming can make up for the shortcomings of digital method to a certain extent, but the phase shifter of antenna unit is controlled by digital, so it still can not reach the ideal simulation state. The hybrid method combines the advantages of the two methods and can achieve the optimal precoding algorithm. In order to obtain low complexity hybrid beamforming, a precoding algorithm based on compressed sensing theory has been proposed [4].

For beamforming, the angle of departure of transmitter and the angle of arrival of receiver are very important. The relevant data information can be obtained to maintain a stable communication link Through the beam tracking technology. The transmitter sends training beam to a predetermined direction, and AOD is obtained by analyzing the received signal. This kind of beam switching can realize the beam collimation in each communication. In the field of low frequency, the development of beam tracking technology is very mature, which uses two-dimensional beamforming, and the other dimension is omnidirectional radiation. Due to the high propagation loss of terahertz wave, the directivity and gain of antenna are higher than those in the low frequency range, so it requires three-dimensional simultaneous beam scanning, so the calculation complexity, time and hardware design are naturally more difficult [6], provided a fast scanning strategy by combining low-frequency and terahertz bands, but this scheme takes up too much spectrum resources. In addition to the method of searching target by beam scanning. Sakr *et al.*, [7]. also proposed that the position of target in the next time slot can be estimated by algorithm, so as to carry out collimation. With the further development of the prediction algorithm, this method will be widely used in terahertz communication.

Path Loss, Noise, Capacity

In wireless communication system, it is very important to establish an effective channel model to maximize terahertz bandwidth allocation and improve spectrum efficiency. In order to achieve efficient wireless communication channel in terahertz band, characteristics of this band must be considered. Terahertz band has high-frequency attenuation, unique reflection and scattering characteristics, and the spatial distribution of the propagation path has a significant impact on the channel. Considering the free space attenuation, molecular absorption and severe weather conditions, the future terahertz wireless communication system will be widely used in short distance indoor communication scenarios. In addition to the ideal weather environment, specified humidity and temperature, remote outdoor communication scenarios are rarely used in practical applications. The common terahertz wave propagation model can be composed of LoS and NLoS. NLoS path includes reflection path, scattering path and diffraction path. he current indoor terahertz channel model research includes deterministic method [8 - 16], statistical method and hybrid method.

Transmission Loss and Noise

Besides the additive white Gaussian noise (AWGN) that is in microwave band, there is another typical noise in terahertz wireless receiver, that is the molecular absorption noise. The molecular absorption noise is caused by the re radiation of some of the absorbed energy by vibrating molecules. The molecular absorption noise is dispersive since different kinds of molecules have different resonance frequencies. The path loss includes spreading loss and absorption loss [17]. Considering the special characteristics of path loss and noise, the network capacity of terahertz communication system with pulse signal is given as following.

In terahertz communication model, the total path loss is defined as the sum of spreading loss (Aspread) and molecular absorption loss (Aabs) [8]:

$$A(f,d) = A_{spread}(f,d) + A_{abs}(f,d) \qquad (1)$$

f is the frequency of terahertz wave and d is the total path length. Propagation loss refers to the attenuation caused by the expansion of wave in medium

$$A_{spread}(f,d) = \left(\frac{4\pi f d}{c}\right)^2 \qquad (2)$$

c is the speed of light in a vacuum. Beer Lambert's law shows the molecular absorption loss in the channel, which relies on the ingredients of the medium, relative humidity, pressure and temperature and so on, which leads to frequency selective fading of broadband signals. The absorption loss of molecules can be expressed as follows

$$A_{abs}(f,d) = e^{k(f)d} \qquad (3)$$

$K(f)$ is the overall absorption coefficient of the medium [18]:

$$k(f) = \sum_{g} \frac{p}{p_0} \frac{T_0}{T} \sigma^g(f) \qquad (4)$$

where p is the indoor standard pressure value, p_0 is the reference pressure, T_0 is the standard temperature, T is the system temperature, and σgf is the absorption cross section. The path length d would effect the loss very much. For $d<1$m, the effect of the molecular absorption is ignorable while for d>1m, it turns severe, as Fig. (**9**) shows. Technologies such as distance-aware physical layer design, reflectarrays and intelligent surfaces are applied to overcome the length problem.

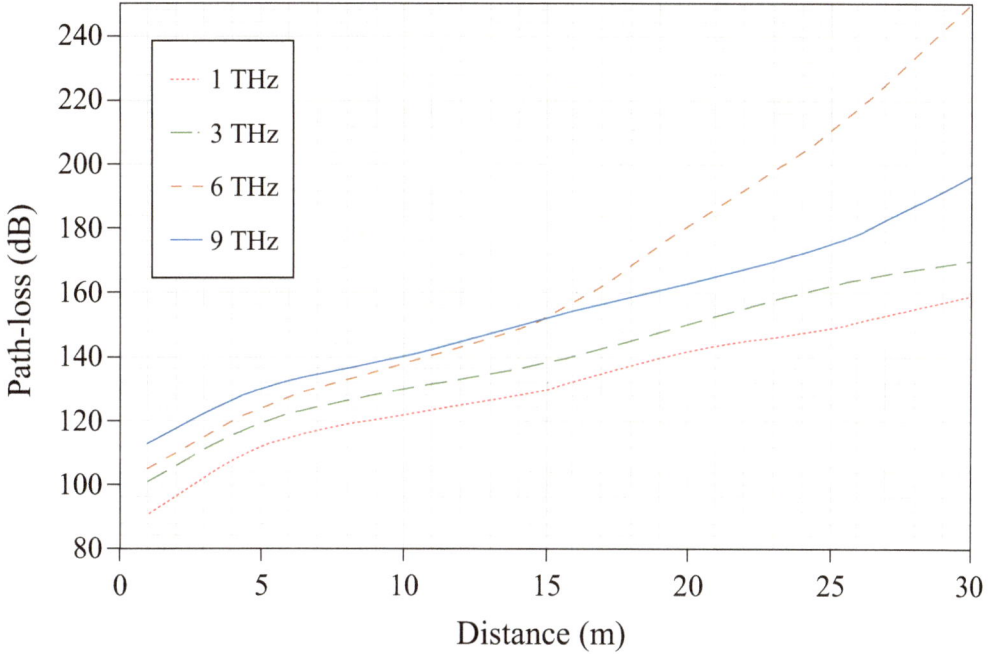

Fig. (9). Path loss varying with distance and frequency.

The environmental noise in terahertz channel mainly comes from molecular noise. The absorption of molecules in the medium not only attenuates the transmitted signal, but also introduces noise. The parameter to measure this phenomenon is the radiation coefficient of the channel, which is defined as:

$$\varepsilon(f,d) = 1 - e^{-k(f)d} \tag{5}$$

Reflection Loss

The reflection loss of terahertz wave is calculated by Kirchhoff theory. The Fresnel reflection coefficient and Rayleigh roughness factor can be used to calculate the reflection loss. The Fresnel reflection coefficient can be denoted as:

$$R(f) = \frac{\cos(\theta_i) - n_t \sqrt{1 - \left(\frac{1}{n_t}\sin(\theta_i)\right)^2}}{\cos(\theta_i) + n_t \sqrt{1 - \left(\frac{1}{n_t}\sin(\theta_i)\right)^2}} \tag{7}$$

where, θ_i is the angle of incidence, n_t is the reflectivity of the medium. The Rayleigh roughness factor is defined as:

$$\rho(f) = e^{-\frac{G(f)}{2}} \tag{8}$$

$$G(f) = \left(\frac{4\pi\omega\cos(\theta_i)}{\lambda}\right)^2 \tag{9}$$

where ω is the standard deviation of surfac6e roughness and λ is the free space wavelength of incident wave. Therefore, the reflection loss can be expressed as:

$$\Gamma(f,d) = \sum A_{spread}(f,d) * A_{abs}(f,d) * R(f) * \rho(f) \tag{10}$$

Scattering Loss

The influence of scattering on terahertz signal is based on the roughness level of the surface. This is considered a critical communication link and must also be considered. Firstly, we consider the calculation theory of Beckman Kirchhoff scattering coefficient:

$$S(f) = -e^{\frac{-2\cos(\theta_1)}{\sqrt{n_{t2}-1}}} * \sqrt{\frac{1}{1 + g + \frac{g^2}{2} + \frac{g^3}{6}}} * \sqrt{p_0^2 + \frac{\pi\cos(\theta_1)}{100} g e^{v_s} + \frac{g^2}{4} e^{\frac{-v_s}{2}}} \tag{11}$$

where P_0 is the renormalization constant, G is the approximate constant, vs is the normalized coordinate coefficient, and θ_1 is the incident zenith angle. Therefore, considering diffusion, molecular absorption, Rayleigh roughness factor and scattering coefficient of N-Ray, the scattering loss can be obtained as follows:

$$\xi(f,d) = \sum A_{spread}(f,d) * A_{abs}(f,d) * R(f) * S(f) \tag{12}$$

Channel Cavity

Terahertz channel has high frequency selectivity and its molecular noise is non-white. Therefore, the capacity can be obtained by dividing the total bandwidth into many narrow subbands and adding the capacities of each subband. The i-th subband is centered on the frequency f_i, $i = 1,2,\ldots$. It has Δf bandwidth. When the sub bandwidth is small enough, the channel appears frequency nonselective, and the noise power spectrum density can be considered as locally flat. The final capacity in bits/sec is:

$$C(d) = \sum_i \Delta f log_2 \left[1 + \frac{B(f_i)A^{-1}(f_i,d)}{N(f_i,d)} \right] \tag{13}$$

B is the transmission signal and N is the noise. In addition to study the specific characteristics of terahertz channel model, it is also worth studying how the observed signal propagates into the medium under the assumption of LOS and NLOS. In fact, there is a very high frequency selective path loss in the Los link, and due to the material type and the roughness of the reflection surface, high reflection will occur in NLOS propagation. The multi ray method can be considered by assuming reflection, scattering and diffraction paths to simulate the main channel characteristics in terahertz band, such as spectral window and time broadening effect with varying distance.

Terahertz Channel Model

Although much attention has been paid to channel modeling in microwave and optical frequency regions, the one in terahertz band still lacks exploration. Recently, terahertz band has received some attention in the literature. According to the application scenarios, the terahertz channel model can be divided into indoor channel model and outdoor channel model.

Outdoor Channel Model

There are few models to simulate terahertz channel in outdoor environment, which only focus on point-to-point link. The first 120GHz experimental radio license and the first outdoor transmission experiment at 170 meters was carried out in 2004. These experiments rely on the use of millimeter wave amplifiers and high gain antennas, such as Gaussian optical lens antenna or Cassegrain antenna. Since 2007, the 120GHz wireless signal is produced by InP HEMT MMIC technology. The advantages of electronic system are compact and low cost. After the introduction of forward error correction (FEC) technology, 5.8 km 10 Gbps data transmission is realized by increasing the output power and antenna gain [19]. By using QPSK modulation, the transmission data rate is further increased to 22.2 Gbps. The current outdoor channel model only deals with point-to-point cases. This is because there are few reports of experimental measurements in the literature. For outdoor measurements, bit error rate (BER) performance can be severely affected by non line of sight paths. For long-distance wireless communication, terahertz link will suffer significant signal loss due to the influence of atmospheric weather. However, despite the presence of absorption peaks concentrated at several specific frequencies, the available transmission window allows the establishment of *via*ble communications in the terahertz band. Therefore, the effect of weather on high-capacity data links is one of the key differentiators between THz links and other wireless methods. Due to the high frequency selectivity of terahertz channel, the transmission distance is limited by attenuation, so the carrier frequency needs to be determined on a case-by-case basis. In order to develop terahertz outdoor channel model, it is necessary to use real data stream to evaluate the link performance. Further exploration of geometry based, visual area based, and map based models including parameterization of measurement results can obtain a complete outdoor channel model. It must be taken into account that in outdoor environments, a number of solutions are needed to eliminate interference from spurious signals in the same frequency band. The first channel characteristic of train to train link is completed at 300 GHz and verified by ray tracing simulator. For the vehicle communication channel, some measurement methods are introduced to measure the multipath reflection including the influence of bypass and multi-lane.

Indoor Channel Model

Different from outdoor channel model, indoor channel model can be divided into deterministic method model, statistical method model and hybrid method model [3].

For deterministic channels, ray tracing model is usually used. Based on the principle of geometrical optics, this method is used to analyze the propagation path of line of sight (LOS) and non line of sight (NLOS). It is site-specific, in line with the propagation theory, and accurately captures the wave propagation phenomenon. However, the accuracy of ray tracing model largely depends on the comprehensive understanding of material properties. Therefore, it is necessary to constantly adapt the model to the new environment, thus limiting its time efficiency. From the perspective of communication, it is very important to understand the large-scale and small-scale statistical data of channel, including path loss, shadow and multipath propagation. Therefore, the statistical method produces a suitable channel measurement based on experience. The first terahertz channel statistical model ranges from 275 GHz to 325 GHz. The model relies on a large number of ray tracing simulations to achieve channel statistical parameters.

Channel statistics such as correlation functions and power delay profiles cannot be easily captured. In order to solve these problems, Kim proposed a geometric statistical model of the device to device (D2D) scattering channel in the Asia Pacific Hertz band. These models simulate the scattering and reflection modes in the D2D environment of Asia Pacific hertz. It must be pointed out that since the reflection and scattering characteristics in terahertz band are frequency dependent, it is necessary to properly simulate the statistical distribution and parameters within and between clusters. Therefore, many studies have considered the characteristics of scattering multipath clusters, including arrival angle and arrival time, in terahertz indoor channel modeling. Choi *et al.* proposed an improved terahertz channel model and a path selection algorithm to find the main signal. Similarly, Kokkoniemi uses stochastic geometric analysis to study the average interference power and the THz band outage probability. In addition, Tsujimura proposed a time domain channel model of terahertz band, in which the coherent bandwidth of the entire terahertz band and its subbands is calculated. Numerical calculation and experimental results show that the obtained impulse response satisfies the causality, and the understanding of the coherent bandwidth variation allows the selection of appropriate center frequency for wireless communication in terahertz band. Different from the traditional channel measurement, scene specific models can also be found in the literature. He proposed a stochastic model of terminal applications in terahertz, especially between 220GHz and 340GHz. The channel characteristics of three different kiosk application scenarios are extracted by 3D ray tracing simulator. In addition, Peng *et al.* proposed a random channel model for future wireless terahertz data centers. The proposed random channel model takes into account the spatiotemporal discreteness of the propagation path, which makes it possible to generate the channel quickly. The root mean square delay and angle spread are used to verify the model.

In order to obtain both high accuracy and high efficiency, Chen *et al.* proposed a chip-to-chip hybrid channel model that combines deterministic and statistical methods [20]. In their discussion, the authors note that a hybrid method of random scatterer placement and ray tracing can be developed. In this case, scatterers are randomly placed, and multipath propagation is tracked and modeled in a deterministic manner based on ray tracing technology. Therefore, a geometric random channel model is established. This hybrid approach has a high modeling accuracy and a simple structure, which allows the use of statistical modeling to obtain rich multipath effects and to calculate the critical multipath component. On a similar frontier, Fu *et al.* demonstrated chip to chip communication by describing the propagation characteristics of computer desktop applications in a 300 GHz metal enclosure. The LOS and NLOS measurements are provided. Compared with the case of free space, the multipath in this case is due to the alternation of traveling waves between the receiver and transmitter of the cavity, which leads to the stronger fluctuation of the path loss, and thus reduces the bandwidth of the channel. In addition, Zhao *et al.* gives channel measurements for 650 GHz carrier and 350 GHz carrier in a typical indoor environment, presents an extensive multi-path channel model that describes the spatial distribution of all available paths and their respective power levels. Therefore, the propagation of terahertz wave in different wavelength range provides a clearer understanding.

In recent years, micro sensing devices (called nanodevices) developed using the properties of new nanomaterials can perform a range of tasks including computing, data storage, sensing and actuation. In addition, the formation of Nanonetworks based on nanodevices will lead to a wide range of applications in biomedical, industrial and military fields. Based on radiation transfer theory and molecular absorption, Joenet proposed a physical channel model for wireless communication between nano devices in terahertz band. The proposed model takes into account the contributions of different types and concentrations of molecules, and HITRAN database is used to calculate the attenuation of waves. According to the beer Lambert law, the transmissivity of the medium dependent on the absorption coefficient of the medium is calculated. The model is also used to calculate the channel capacity of Nanonetworks operating in terahertz band. In this case, different power allocation schemes are deployed. In order to ensure the strong signal received, the author suggests to use a lower absorption coefficient in terahertz band. In addition, the sky noise model is the basis of the existing absorption noise models. Kokkoniemi elaborated on this topic and put forward different views on how to simulate molecular absorption noise. However, there is no actual experiment to verify the proposed model. Not only the absorption, but also the scattering of molecules and small particles also affect the propagation of electromagnetic waves. Therefore, the Kokkoniemi scattering model is used to demonstrate the frequency of the broadband terahertz. In addition, an omnidirectional geometric model based on Petrov is proposed to deal with the interference. However, in their model, they ignore the interference caused by the presence of base stations. Wang solved this problem, and they studied the interference of beamforming base stations [21]. Therefore, it is recommended to use high-density base stations, use beamforming with small beam width antennas and deploy low-density nano sensors to improve coverage probability.

Ray Tracing

Ray tracing has become a popular technique for analyzing site-specific scenarios because it can analyze very large structures with reasonable computational complexity.

As shown in Fig. (**10**), radio signals transmitted from a fixed light source encounter multiple objects in the environment, forming line of sight, reflection, diffraction or scattering of light reaching the receiver. For each geometric path, ray tracing technology analyzes the propagation of electromagnetic waves by representing the wavefront as simple particles.

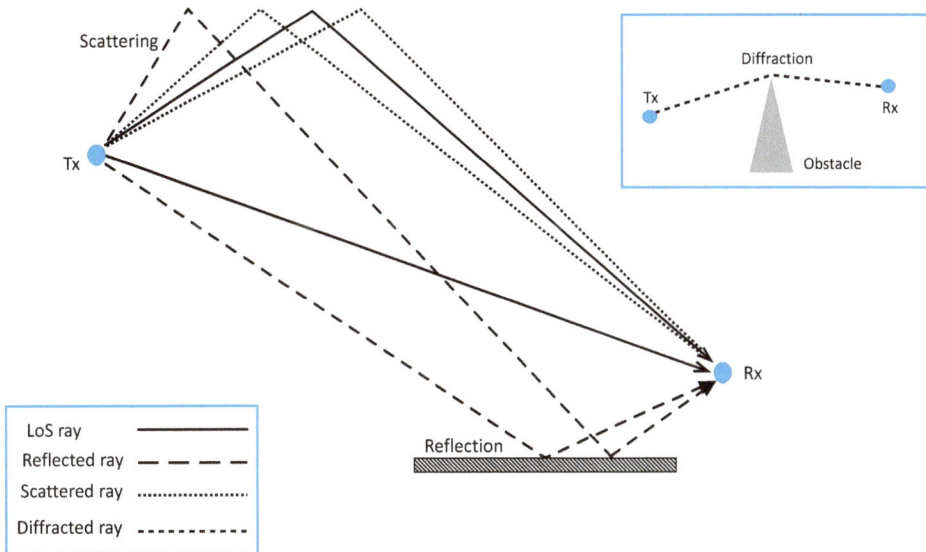

Fig. (10). Schematic diagram of ray propagation.

In terahertz band, free space propagation needs to consider the propagation loss of Frith's law and the molecular absorption loss caused by the conversion of part of wave energy into molecular internal kinetic energy. In addition, the reflection, diffraction and scattering effects are similar to geometrical optics because they have stronger particle properties at terahertz and require more accurate calculations. Sheikh proposed a self-programming 3D ray tracing algorithm based on Beckmann Kirchhoff model to simulate the diffuse scattering mechanism in non reflective direction at terahertz frequency [22]. In the spatial and temporal dispersion of the submillimeter wavelength band, the incident light may split into a specular and several non-specular (diffusely scattered) rays after bouncing off a rough material, which accounts for a large proportion of all propagation light. Therefore, propagation modeling must be considered to accurately predict channel characteristics. Han proposed a unified multi-ray channel model based on bottom-up method in 0.06-10 terahertz band.

Terahertz Wireless Mobile Communications

The terahertz wireless communication is expected to have a data rate of multiple tens of Gbps or even several Tbps. It can be predicted that terahertz communication will have important application prospects in many aspects.

WPAN and WLAN

the WPAN systems have the potential to realize specific applications such as mobile phones, earphones and other tablet-like devices, as Fig. (**11**) shows. The access points (APs) may be developed in some places like shopping mall halls, metro station gates and other indoor places with high human mobility [4]. Compared with the AP that adopts the low frequency band, one of the most prominent characteristics is that the AP based on THz band provides the ability to simultaneously transmit information to multiuser in different directions by using the sub-array antenna structure. With the further development of the AP hardware devices, the AP equipment is available to bring the excellent communication service to human life and facilitates the high-quality video in offices and living rooms, *e.g.*, the HD holographic video conferences, the ultra-high resolution video formats, the download of HD film files and the VR technique [4].

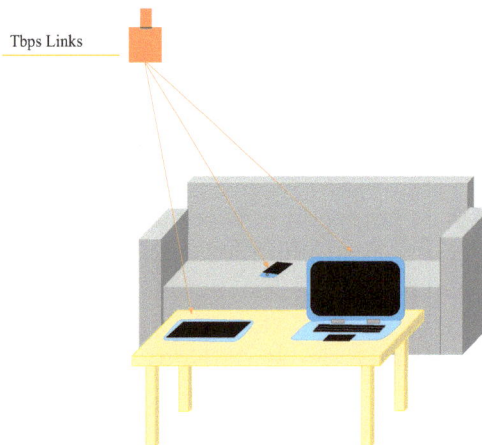

Tbps Links

Fig. (11). Terahertz application scenarios for WPAN.

VR

It is worth emphasizing that the development of the VR technique is heavily limited by the low-data-speed wireless communication at present. Fortunately, once the THz bandwidth is employed in wireless communication systems, the VR technique will bring much better user experience than the wire communication systems.

Augmented reality and virtual reality (AR/VR) are considered to be one of the most important requirements of 5G, especially one of the typical applications with high throughput requirements. 5G will be able to support the transformation of current wired or fixed wireless access AR/VR into wireless mobile AR/VR in more extensive scenarios, as shown in Fig. (12). Once AR / VR can be used more easily and freely, it will promote the rapid development of AR/VR business. The media interaction form is mainly based on the current plane multimedia. It is foreseen that in the era of 6G, it will develop into high-fidelity AR/VR interaction and even holographic information interaction, where holographic communication can take place anytime and anywhere to achieve the communication vision of "holographic connection", and the high transmission rate provided by terahertz communication is a reliable choice to realize this vision.

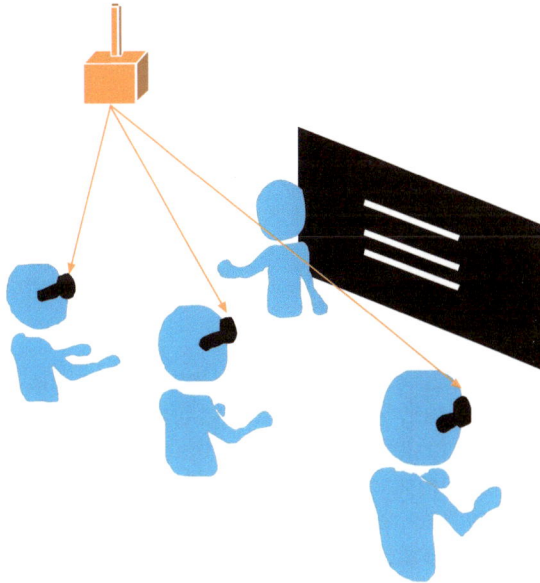

Fig. (12). Terahertz application for VR.

Directional Networks

The beam in terahertz band is narrower than that in lower frequency band, and the coverage area of terahertz beam is limited. The current omni-directional networking technology based on omni-directional antenna is hard to discover network nodes quickly, complete directional networking and save system energy consumption. Different from omni-directional antenna, the use of directional antenna in wireless communication system can bring significant performance improvement to the network. Directional antenna can concentrate the energy of terahertz wave in a specific direction to send information, which is usually called the main lobe. In addition, when the directional antenna is used to realize directional networking, the interference between these neighbor nodes is greatly reduced, which enables multiple pairs of neighbor nodes in the network to send messages to each other and greatly increases the network capacity. Finally, for the communication system with fixed transmission power, the directional antenna can greatly increase the transmission distance.

Secure Communications

Terahertz system will use directional narrow beam, which can not only expand the communication range, but also partially protect the data in the physical layer. In most cases, the attacker must be in the coverage of transmitter beam in order to eavesdrop on the message. While even using a very narrow beam would result in a considerable area around the receiver where the attacker can capture all the data. The physical layer of millimeter-wave and terahertz band communications utilizes the inherent properties of a directionally narrow beam to greatly enhance security. The reason is that the attacker must be physically in the coverage of transmission beam to decode the data. Using narrow beam and physical layer security specific coding can greatly reduce the probability of data eavesdropping under LOS and NLOS conditions.

Application scenarios of secure communication system are military and defense fields. In the field of military confrontation, UAV can replace human pilots to perform some dangerous tasks, such as combat, reconnaissance, *etc*. When the UAV completes the reconnaissance mission, the undistorted high-definition video information obtained by the reconnaissance UAV can be transmitted to other combat units, such as unmanned reconnaissance aircraft, manned aircraft, tanks, *etc*., and the combat units can better analyze the combat environment. In addition, terahertz band has unique advantages in security communication system, so it is difficult for enemy combat units to intercept military intelligence from very narrow terahertz beam. Due to the ultra-high bandwidth of terahertz, spread spectrum technology and frequency hopping technology can be used to combat interference attacks [4].

SUMMARY

Owing to its high data rate, terahertz technology is a strong candidate for 6G communication. While the following aspects need to be further studied in order to support terahertz communication: Semiconductor technology, including RF, analog baseband and digital logic, *etc*; High speed baseband signal processing technology and integrated circuit design method with low complexity and low power consumption, terahertz high-speed communication baseband platform; Modulation and demodulation, including terahertz direct modulation, terahertz mixing modulation and terahertz photoelectric modulation; Waveform and channel coding; Synchronization mechanism, such as high-speed and high-precision acquisition and tracking mechanism, synchronization mechanism of hundreds of orders of magnitude antenna array. Besides, other potential research topics related to terahertz communication may be channel modeling, basic system concepts and some technology-related research such as terahertz transmitters and mixers [23].

To use the terahertz wave in 6G, the most urgent hardware of IoT/mobile devices in need is perhaps battery with large capacity, due to the high path loss of terahertz wave propagating in atmosphere which in turn requires high transmit power. Beamformers and antennas with high gain are also in need for the same reason. Meanwhile, as the photonics-defined radio system is the natural evolution to 6G, photonic hardware, *e.g.* a photonic RF front end are also in need.

Research on terahertz communication technology has only been two decades, many key devices have not been developed successfully, and some key technologies are not mature enough, and a lot of research work is needed. However, terahertz communication is a promising technology, With the breakthrough of key devices and technologies, terahertz wave communication technology will have a profound impact on human production and life.

CONSENT FOR PUBLICATION

Not applicable.

CONFLICT OF INTEREST

The author declares no conflict of interest, financial or otherwise.

ACKNOWLEDGEMENTS

Declared none.

REFERENCES

[1]   E.C. Strinati, "6G: the next Frontier from holographic messaging to artificial intelligence using subterahertz and visible light communication", *IEEE Veh. Technol. Mag.*, vol. 14, pp. 42-50, 2019.
[http://dx.doi.org/10.1109/MVT.2019.2921162]

[2]   K.M.S. Huq, "Terahertz-enabled wireless system for beyond-5G ultra-fast networks: a brief survey", *IEEE Netw.*, vol. 33, pp. 89-95, 2019.
[http://dx.doi.org/10.1109/MNET.2019.1800430]

[3]   H. Elayan, "Terahertz band: the last piece of RF spectrum puzzle for communication systems", *IEEE Open Journal of the Communications Society*, vol. 1, pp. 1-32, 2019.
[http://dx.doi.org/10.1109/OJCOMS.2019.2953633]

[4]   Z. Chen, "A survey on terahertz communications", *China Commun.*, vol. 16, pp. 1-35, 2019.
[http://dx.doi.org/10.23919/JCC.2019.09.001]

[5]   Y. Ma, J. Grant, S. Saha, and D.R. Cumming, "Terahertz single pixel imaging based on a Nipkow disk", *Opt. Lett.*, vol. 37, no. 9, pp. 1484-1486, 2012.
[http://dx.doi.org/10.1364/OL.37.001484] [PMID: 22555712]

[6]   B. Peng, S. Priebe, and T. Kurner, "Fast beam searching concept for indoor terahertz communications", *European Conference on Antennas & Propagation*, 2014pp. 639-643
[http://dx.doi.org/10.1109/EuCAP.2014.6901840]

[7]   X. Gao, "Fast channel tracking for terahertz beamspace massive MIMO systems", *IEEE Trans. Vehicular Technol.*, vol. 66, pp. 5689-5696, 2017.
[http://dx.doi.org/10.1109/TVT.2016.2614994]

[8]   J.M. Jornet, and I.F. Akyildiz, "Channel modeling and capacity analysis for electromagnetic wireless nanonetworks in the Terahertz band", *IEEE Trans. Wirel. Commun.*, vol. 10, pp. 3211-3221, 2011.
[http://dx.doi.org/10.1109/TWC.2011.081011.100545]

[9]   C. Han, A.O. Bicen, and I.F. Akyildiz, "Multi-ray channel modeling and wideband characterization for wireless communications in the terahertz band", *IEEE Trans. Wirel. Commun.*, vol. 14, pp. 2402-2412, 2015.
[http://dx.doi.org/10.1109/TWC.2014.2386335]

[10]   A. Moldovan, ""LOS and NLOS channel modeling for terahertz wireless communication with scattered rays", in 2014", *IEEE Globecom Workshops*, pp. 388-392, 2014.

[11]   J. Kokkoniemi, "Frequency and time domain channel models for nanonetworks in terahertz band", *IEEE Trans. Antenn. Propag.*, vol. 63, pp. 678-691, 2015.
[http://dx.doi.org/10.1109/TAP.2014.2373371]

[12]   S. Priebe, "A comparison of indoor channel measurements and ray tracing simulations at 300 GHz", *35th International Conference on Infrared, Millimeter, and Terahertz Waves*, 2010
[http://dx.doi.org/10.1109/ICIMW.2010.5612330]

[13]   A. Gureev, "Channel description in the low-THz wireless communications", *2017 IEEE Conference of Russian Young Researchers in Electrical and Electronic Engineering*, 2017pp. 1240-1243
[http://dx.doi.org/10.1109/EIConRus.2017.7910787]

[14]   C. Han, and I.F. Akyildiz, "Three-dimensional end-to-end modeling and analysis for graphene-enabled terahertz band communications", *IEEE Trans. Vehicular Technol.*, vol. 66, pp. 5626-5634, 2017.
[http://dx.doi.org/10.1109/TVT.2016.2614335]

[15]   C. Zhang, C. Han, and I.F. Akyildiz, "Three dimensional end-to-end modeling and directivity analysis

for graphene-based antennas in the terahertz band", *2015 IEEE Global Communications Conference,* 2015
[http://dx.doi.org/10.1109/GLOCOM.2015.7417131]

[16]   D. He, "Channel modeling for Kiosk downloading communication system at 300 GHz",
[http://dx.doi.org/10.23919/EuCAP.2017.7928447]

[17]   L. Zhang, Y.C. Liang, and D. Niyato, "6G visions: mobile ultra-broadband, super internet-of-things, and artificial intelligence", *China Commun.,* vol. 16, pp. 1-14, 2019.
[http://dx.doi.org/10.23919/JCC.2019.08.001]

[18]   M. Pengnoo, "Digital Twin for Metasurface Reflector Management in 6G Terahertz Communications", *IEEE Access,* vol. 8, pp. 114580-114596, 2020.
[http://dx.doi.org/10.1109/ACCESS.2020.3003734]

[19]   A. Hirata, "120-GHz-band wireless link technologies for outdoor 10-Gbit/s data transmission", *IEEE Trans. Microw. Theory Tech.,* vol. 60, pp. 881-895, 2012.
[http://dx.doi.org/10.1109/TMTT.2011.2178256]

[20]   Y. Chen, and C. Han, "Channel modeling and analysis for wireless networks-on-chip communications in the millimeter wave and terahertz bands", *IEEE Conference on Computer Communications Workshops,* 2018pp. 651-656
[http://dx.doi.org/10.1109/INFCOMW.2018.8406954]

[21]   C.C. Wang, "Interference and coverage analysis for terahertz band communication in nanonetworks", *IEEE Global Communications Conference,* 2017
[http://dx.doi.org/10.1109/GLOCOM.2017.8255059]

[22]   F. Sheikh, D. Lessy, and T. Kaiser, "A novel ray-tracing algorithm for non-specular diffuse scattered rays at terahertz frequencies", *2018 First International Workshop on Mobile Terahertz Systems,* 2018
[http://dx.doi.org/10.1109/IWMTS.2018.8454694]

[23]   T. Kürner, and S. Priebe, "Towards THz communications - status in research, standardization and regulation", *J. Infrared Millim. Terahertz Waves,* vol. 35, pp. 53-62, 2014.
[http://dx.doi.org/10.1007/s10762-013-0014-3]

CHAPTER 5

Intelligent Network Slicing Management and Control for 6G Mobile Networks

Fanqin Zhou[1,*] and **Mohamed Cheriet**[2]

[1] *State Key Laboratory of Networking and Switching Technology, Beijing University of Posts and Telecommunications, Beijing, P. R. China*

[2] *École de Technologie Supérieure, Université du Quebec, Montreal, Canada*

Abstract: The Internet of Things (IoT) is a key enabler of smart cities, where a variety of applications proliferate to help citizen services. As different IoT applications have different service holders, it becomes necessary to employ network slicing (NS) to gain distinct virtual networks, and differentiated quality of service (QoS) guarantees. Other than conventional IoT scenarios, smart city IoT relies on 6G networks for broad coverage, ultra-low latency, and reliable connection. This chapter proposes a self-organizing network (SON) driven network slicing architecture, where software-defined networking (SDN) and network function virtualization (NFV) also play important roles. Some preliminary simulation results are given to validate the efficiency of our design.

Keywords: Intelligent Network Slicing, Internet of Things, Self-organizing Network, Software-defined Networking.

INTRODUCTION

Different from 5G networks that are designed for digitalizing several urgent scenarios in modern society, such as massive connection, enhanced mobile broadband, and ultra-low latency, networks in the 6G era are envisioned to be full dimensional networks that would address every potential demand of network services. For example, a smart city is a promising solution to improve citizens' quality of life. However, it will heavily rely on 6G mobile networks, especially the mobile Internet of things (IoT), to connected utility infrastructure, public assets, *etc.*, throughout the city to endow the governors and utility service providers with sharper insights to better their services [1]. These IoT devices may come for different utilization purposes, such as surveillance, metering, actuating,

* **Corresponding author Fanqin Zhou:** State Key Laboratory of Networking and Switching Technology, Beijing University of Posts and Telecommunications, Beijing, P. R. China; Tel: 0086 10 62283295; Fax: 0086 10 62283255; E-mail: fqzhou2012@bupt.edu.cn

and from different utility service providers and public sectors, posing a wide range of preferences on the performance of network services. This is quite different from 5G, where services are roughly categorized into three types, namely massive machine type communication (mMTC), ultra-reliable low-latency communication (URLLC), and enhanced mobile broadband (eMBB).

How to serve network applications with extremely different performance requirements without impacting each other was once a critical issue when designing 5G, and network slicing (NS), the concept borrowed from computer network virtualization technologies [2], was thus introduced to address this issue. Through network slicing, 5G mobile network with nation-wide coverage, would be sliced into logically independent layers or parts, or network slice instances (NSIs) specifically. Network services with similar properties will be gathered into an NSI, which usually is allocated a predefined set of resources to keep from the impact of other services. Data flows are arranged in the slice granularity, so the reduced scheduling complexity and cost-effective resource consumption can be achieved.

Due to the logical independence, or isolation specifically, network slicing grants mobile network operators (NOPs) the valuable opportunity to stretch their business to public utilities and vertical industries. In the past, companies in these industries prefer to build private networks for business privacy and designable QoS policies. However, the new wave of society digitization is forcing these companies to extend their perception edge to places where their self-build private networks are hardly accessible. Further building private networks means utility service providers have to pose their network facilities out from their private domain into off-premises environments. Maintaining massive facilities in a huge metro or even nation-wide scope is challenging. Moreover, computational complex tasks, such as augmented reality, video data analyzing, will be a necessary part of future IoT applications, which require sufficient computational and storage resources available as close to the scene as possible. Budget, technology backups, privacy, and performance issues limit further expansion of self-built private networks. Instead, renting slices from mobile network operators becomes an attracting solution. Thus, the properties and types of network services is going to experience a boom before we are entering 6G era.

To satisfy the broad demands on customized IoT networks of smart city service providers, it is foreseeable that a lot of NSIs will be subscribed by different USPs (tenants in NOP's view) simultaneously operating on the same network infrastructure, and each NSI is attached with an appointed service level agreement. How to optimally allocate various network resources to slice instances on demand turns one of the key problems in network slicing. The process, comprised of allocating heterogeneous network resources to NSIs and coordinating the allocations between them, should be flexible to the instant requests from tenants, which by no means can be manually accomplished. To facilitate the resource management process for network slicing, it is necessary to introduce automation properties into the process and frame it into a properly designed network slicing management architecture.

This chapter applies self-organizing approaches in 5G network slicing for smart city IoT. Specifically, a self-organizing network (SON) driven slicing management architecture framework is proposed, which consists of properties like self-organizing, traffic-aware, and robust-guaranteeing to make good utilization of network facilities. The functional framework and management process flow are characterized. A case study of network resource optimization for delay-sensitive (DS) slices and non-delay-sensitive (nDS) slices is given at last to exemplify the effect of flexible resources allocation, together with some analyses on the numeric results.

NETWORK SLICING AND ITS NEW REQUIREMENTS FOR 6G

This section first presents an illustration of network slicing for smart city scenarios with diverse IoT applications, and then introduces a general network slicing management architecture as well as the resource management issues during slicing.

The Concept of Network Slicing

The concept of network slicing is so vivid that it is not hard to imagine the key idea of the technology is to slice the same network infrastructure into several independent logical networks [3]. The way a network slice instance being implemented can be static or dynamic [4]. As to static slicing, network resources are reserved exclusively for individual network slice instances. Thus, static slicing can be easily implemented and naturally makes network services within a slice instance isolated from other network services, while it is obvious that static slicing has the disadvantages of underutilization of network resources. For dynamic slicing, resources are allocated dynamically according to the actual needs of services in the slice instances which is of course beneficial to improve the network resource utilization efficiency, but additional mechanisms are needed to form QoS guarantee of end-to-end (E2E) performance. Thus, both types of slices will be running in future 6G networks, but dynamic slicing contributes more complexity. Thus, a critical goal of network slicing management is to achieve dynamic slicing.

Network Slicing Management in 5G

In 5G networks, a consensus has been reached that the full-fledged network slicing will be based on a variety of the enabling technologies, especially SDN and NFV [5], and a proper network slicing management architecture becomes a necessity to coordinate different technologies and resources. There is not yet a standardized network slicing management architecture for dynamic slicing, whereas existing research outcomes have some features in common. One instance of the typical network slicing management architectures is shown in Fig. **1**. In the architecture, the component functions can be grouped into two types for slice implementation and slice management, respectively.

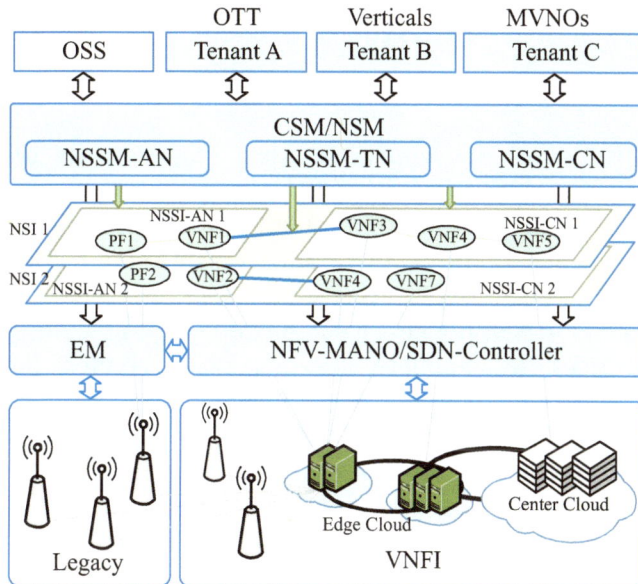

Fig. (1). A typical network slicing management architecture.

The slice management functions directly serve the life-cycle management of network slice instances, which necessarily include the function of network slice management (NSM) with the supports from network slice subnet management (NSSM) functions, managing the subnet slice instances in the subnetwork domains, like access network (AN), transport network (TN), and core network (CN). The slice implementation functions, including legacy entity management (EM), SDN controller, and NFV management and orchestration (MANO), are mainly to compose physical or virtual network functions into a chain of services according to a slice request.

When a communications request from a network service tenant arrives, CSM interprets it to the network slice request. NSM is usually with good knowledge of how a series of network functions will logically form a network service in the form of service function chains (SFCs) to serve the need of a particular tenant, such as over-the-top service provider, vertical industry company, or mobile virtual network operators. It translates the network slice request into SFC and then divides the SFC into segments for subnets for implementation convenience. Then, NFV-MANO, SDN controller and EM cooperate with each other to implement slicing by instantiating the SFC.

To implement network slicing, one of the key issues is the flexible management of cross-domain heterogenous network resources. In the following part, we will briefly discuss the managed resources in network slices in different network domains.

1. Radio access network resources: the main resources in this domain include radio carrier, spectrum band, physical resource block, as well as the radio access network facilities, *e.g.*, antenna, cell sites [6], *etc.* New radio technologies, such as numerology option, transmission mode selection, provide new resource dimensions [7]. According to required resource isolation level and isolation cost RAN, resources can be dedicated or shared [8], and specially reserved resources are wanted by mission-critical services because they permit isolation and stable performance. However, to improve the utilization of dedicated resources, it is necessary to share the radio access network resources flexibly [9]. In terms of shared resource management, it can be achieved by modifying the scheduler to serve users in different slices with the same pool of available resources [2]. However, sharing resources usually implies enforced isolation between slices is not guaranteed, and traffic changes may cause one slice to affect another. Thus, slice-ware dynamic scheduling is needed.

2. Transport network resources: In this domain, the most important resources for slicing are available links and the maximum bandwidth for each link. In some prior work, the server resources attached to TN nodes are also viewed as a part of TN resource. However, this article views these nodes as distributed cloud nodes in CN. On TN resources management, literature [10] studied the next-generation transmission network architecture of forwarding and backhaul fusion and designed a multi-tenant control plane allowing the flexible allocation of transmission resources and cloud resources. In existing works, transmission resources generally form a set of important constraints that determine the VNFs embedding, whereas the problem of determining the link bandwidth between VNFs is NP-hard [11]. This forms a continuous challenge

for network slicing, where computational intelligence approaches will find more applications.

3. Core network resources: In the SDN/NFV enabled network scenarios, core network can be regarded as linked core network function operating on different types of cloud infrastructure. Thus, we say the resources used for slicing in the core network domain are node resources in edge and core cloud, typically including computing, networking, storage/cache resources. The resource allocation may be partially or fully constrained link bandwidth between nodes. Literature [12] investigated the cache resource allocation and designed a chemical-reaction-based algorithm with relaxed by link constraints for efficiency, while a similar core network resource allocation is modeled as a virtual network embedding problem with location constraints and a compatibility-graph based algorithm is proposed to solve the mapping problem between nodes and links [13].

As shown above, various resources in different network domains should be used and coordinated when network slice instances are built. Facing greatly varied 6G network services, to make those slice instances swiftly adaptive to the real-time service demand, dynamic slicing with automatic features is in great need. To this end, we propose a SON-driven network slicing management architecture, show its main self-organizing control loops, and identity three reliable resource management approaches for dynamic slicing.

Requirements of 6G Network Slicing

This section focuses on the requirements for 6G network slicing, as shown in Fig. 2. Other than the dynamic slicing requirement inherited from 5G, other requirements include flexibility and scalability in the slicing management architecture, efficient resource allocation and orchestration, adaptive service function chaining (SFC), and recursion [14].

Flexibility and Scalability in the Slicing Management Architecture

Frequent addition, removal, and swiftly mobility of nodes can be seen in the emerging network environments, such as smart cities, smart industries, and intelligent transport. The dynamic changes of network topology usually cause obvious QoS loss for network services in those smart environments, and the effects will be more prominent in the future 6G era. Besides, other than terrestrial networks, 6G networks are envisioned to be composed of other segments, such as space, aerial, and marine segments. On one hand, these segments are with different features, which should be considered in 6G network slicing, so that it's a particular set of models and abstractions will be used to characterize each segment. On the other hand, the ever-increased network scale due to the integration of different segments is making it increasingly difficult to manage the whole network with a single management entity in a centralized manner. Though earlier networks benefited from the centralized management in aspects such as implementation simplicity; however, when the scale of networks is getting larger, the use of a single management entity following the centralized manner just increases the delay, produces large communication overhead, and adds additional complexity [15]. In the upcoming 6G, networks are envisioned to have a variety of segments to support massive and ubiquitous machine-type connections even attached with ultra-reliable low latency communication and/or wideband requirement, the problems on efficient network management and network-based service provision will be more prominent and challenging [16].

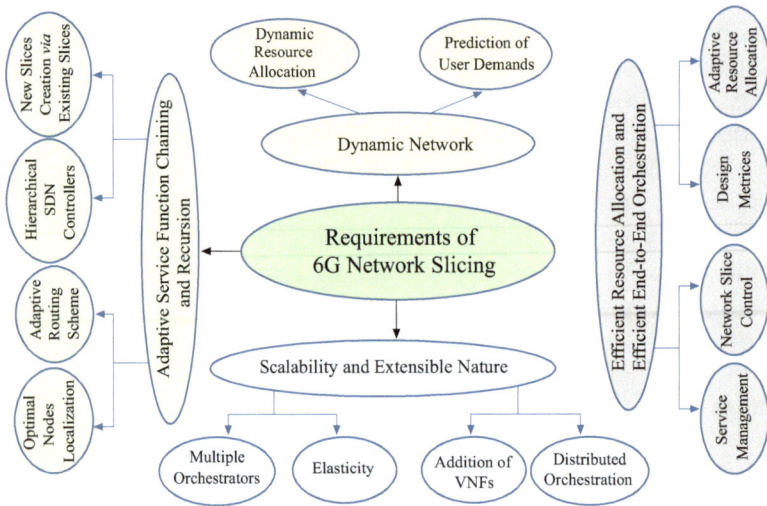

Fig. (2). Network slicing enabled smart city requirements.

In such context, the way how to design, deploy, operate, and upgrade networks need a radical rethink in terms of required resources, temporal delay, and functional flexibility to support the novel network applications with extreme requirements. The network slicing concept has been introduced in 5G and the pros and cons of static and dynamic slicing have been widely discussed. To support the massive QoS-demanding and ever-emerging B5G/6G network applications, network slicing should be capable of extending its functionalities and the overall network resource allocation should be controlled carefully and efficiently. The hierarchical orchestrators-based architecture is regarded as a promising way. In the architecture, multiple orchestrators with each one designed to control a particular network segment are utilized to reduce the complexity. These multiple orchestrators are then controlled by higher-level orchestrators. Thus, lower-level orchestrators execute the responsibility by providing a group of primarily composed virtual network functions to the next higher-level orchestrators to support the scalable operation. Through elastically handling network slicing to manage the allocation of total network resources among segments, underutilization of resources can be alleviated to accommodate more users by the system.

Efficient Resource Allocation and Orchestration

6G networks are designed to support various intelligent services. A network slice consists of different types of resources, which can be combined in an appropriate way to meet the requirements of intelligent services supported by such a slice [5]. These intelligent services impose strict restrictions on the network slicing since all resources have limitations. In particular, the network slices are expected to be isolated, which means a slice should have little impact on other slices. If there is any resource state change in a slice, such as a network flow, the allocated resources of other slices should not be affected. Besides, resource management of each slice should be independent, which means it is permissible for each slice to implement a different access control strategy. Furthermore, resource utilization should be as efficient as possible in order to increase the capacity of base stations and make full use of spectrum resources [17].

Efficient allocation of the spectrum, bandwidth and computing resources of network slices is important to realize joint optimization of the user performance and operating income. Resources are manageable units, which are described with a collection of properties or capacity metrics used for service delivery. In network slicing, two kinds of resources are considered, Network Functions (NFs) and Infrastructure Resources (IRs). NFs are functional units providing particular network functions to support specific services required by each user. NFs are usually software instances that run on IRs. Network functions can be physical, integrating vendor-specific hardware and software, and defining a dedicated physical facility. They can also be virtualized, since the NF software is separated from the hardware IRs consist of fundamental hardware and software used to host and connect NFs, which comprise computing, memory and communication resources, and physical components used for radio access. For ease of utilization in network slicing, it is designed to abstract and logically divide resources and their properties through a virtualization scheme, and virtual resources can be used just like physical resources. To achieve effective network slicing, the requirements for slicing must be accurately computed, and then resource allocation must be performed. The slice resource allocation should be adaptive to ensure QoS. For example, smart transportation is latency-sensitive, so smart transportation slices require more resources to fulfill the strict latency constraints.

The network slicing architecture is composed of a service creation layer that creates services and a network slice management layer that manages resources [18]. The aforementioned processes ensure the efficient management of virtualized network functions and infrastructures for slices with various features. Slice creation is a necessary step to ensure that services are isolated from each other. After receiving a request to create a slice, the control unit will admit slicing if there are enough resources, and a brand-new slice will be generated by the slice template and network function instance. Then, resources should be orchestrated properly and allocated to slices to satisfy their service level agreements. Because of multi-dimensional resource requirements for real-time services, a slice will be assigned multiple virtualized resources, which are mapped to physical network infrastructures through resource mapping algorithms [18]. An efficient architecture for network slicing orchestration is indispensable to effectively manage network slicing. Literature [19] designed an architecture for network slicing orchestration which is composed of a service receptor, a virtualized resource orchestrator, a network slice lifecycle manager, and a resource programmable controller. E2E is a built-in feature of network slicing, which helps service providers to deliver services to users. It spans different management scopes, which means the slice is capable of merging resources provided by the different infrastructure. Besides, it can integrate diverse network layers and technologies, such as access network, core network, transmission network, and cloud centers. In particular, various resources supporting an overlaid service layer are allocated to a network slicing, which makes it possible to integrate network and service efficiently. Besides, slice overbooking can provide higher profit for mobile operators. In [20], a hierarchical control plane was proposed to effectively manage network slices, and the slice orchestration problem was modeled as a stochastic yield management problem and solved by two approaches. Experimental results indicated the slice overbooking brought noticeable profit growth. Additionally, artificial intelligence would be a useful tool to facilitate slice overbooking.

Adaptive Service Function Chaining and Recursion

To serve network services with diverse requirements, network slicing architecture for mobile networks are envisioned to be built based on SDN/NFV technologies. Thus, the network slice instances are implemented using chained virtualized network service functions. This design gains networks a new granularity of freedom to adaptively allocate a set of resources to a target service. As each virtual function should be running on a specific node and thus occupies a certain amount of resources within the node, the cost of running the function, such as network latency and operational expenditure, differs from service to service because of the different initiation and termination locations of the services.

Therefore, it is necessary to optimize the location of nodes to realize the adaptive SFC. In 6G, the flexible network architecture also grants the network the capability of supporting the small group-oriented services, such as emergency communication at the scene of fire, crash, or disaster. In these cases, similar services but for different tenants or event scenes will call for additional slices. New slices can be created from existing slices, which is referred to as recursion, instead of generating an entirely new one because creating a new slice is more complex than recursively creating a slice.

To achieve adaptive SFC, optimization of the overall performance of the networks also forms a core requirement, and two main challenges are the selection of the embedding nodes for the virtual functions and the optimal routing schemes among those selected nodes. To address the problem, the joint optimization of link traffic and operating cost is formulated, and a matching theory-based approach is proposed [21]. In [14], the minimization of network traffic is designed to formulate the SFC problem, and a penalty successive upper bound minimization algorithm and its transforms are proposed. The extreme case of adaptive SFC is the recursion. In addition to the resource allocation consideration, recursion from existing slice also raises requirements on the flexibility and efficiency of network slicing management architecture. The hierarchical multiple controller architecture allows controllers to create new slice instances for their clients using resource and context information from other controllers.

SON-DRIVEN NETWORK SLICING FOR 6G NETWORKS

SON-driven Network Slicing Management Architecture

The proposed SON-driven network slicing management architecture is shown in Fig. (**3b**). The traditional concept of a self-organizing network is extended from merely radio access network domain to the whole network, including mobile network core and transport network explicitly, as the self-organizing feature will be a necessity for controlling the complexity in performing dynamic slicing. Besides, SON entities are logical functions that can be implemented as a part of existing management entities or independent function entities, quite flexible in the SDN/NFV context.

In the top layer, NSM SON is proposed to keep surveillance of the E2E performance of different slice instances, takes evaluations of the status, and predictions of the future trend. When a triggering event occurs, without human involvement, it will execute corresponding actions including sending high-level instructions to lower layer SON entities, such as NSSM-AN SON and NSSM-CN SON. As their names indicate, they are responsible for access network and core network, respectively. Due to the close relationship between transport and core networks, NSSM-CN SON takes charge of the self-organizing of both the two network domains.

In the radio access network domain, traditional SON entities still play significant roles. Distributive SON (D-SON) runs in each node to perform automatic resource allocation, energy-saving, load balancing, *etc.*, while centralized SON (CSON) coordinates the behaviors of nodes when performing optimizations. Both types of them need enhancement on slice instance awareness, so that cellular node will be able to schedule radio network resources in user granularity while achieving slicing. Though the enhanced SONs enable nodes to dynamically schedule pooled resources for slices, it is not equivalent to dynamic slicing. To this end, we propose NSSMAN, SON, to responsible for dynamic slicing by self-optimizing the quotas of resources among slices.

In the core network domain, the NFV-MANO and SDN controller cooperate to perform automated network slicing according to the given SFC and the attached service level agreement. During the process, some self-organizing features can also be observed under the control of NFV-MANO and SDN controller, such as substituting mal-functional VNFs and reroute traffic flows from a downed port. However, though having the potential to support dynamic slicing, NFV-MANO and SDN controller are designed to guarantee the quality of service of instantiated SFCs, rather than directly responsible for traffic-aware slice instance tuning with the risk of deteriorating the QoS. Therefore, the NSSM-CN SON is proposed to implement traffic awareness and generate high-level tuning instructions for NFV-MANO and SDN controllers.

Management Process in SON-driven NSMA

As shown in Fig. (**3b**), the SON-driven network slicing architecture has closed control loops in each domain for intradomain self-organizing. Higher layer management functions associate each domain and form a global control loop monitoring overall network resource utilization and performance.

A closed SON control loop typically consists of four phases, *i.e.*, Monitoring, Analyzing, Planning, and Executing, which are abbreviated as M, A, P, and E in the figure. A SON-driven management entity can adopt data analyzing and computational intelligent approaches. Therefore, in the analyzing phase, the management entity can analyze the data intelligently gathered in the monitoring phase to find out the cause of an outlier event and then decide the best action to deal with the case with optimized parameters in the planning phase. After performing the action in the executing phase, it goes back to the monitoring phase and closes the loop.

The SON control loops can be observed in each subnet domain and different management layers. In the radio access network domain, cellular nodes with slice-aware D-SON functions will keep monitoring the local resource utilization rate in different slices, and enter the SON loop when triggering events occur, *e.g.*, a low data rate of terminal devices in a network slice. The node will analyze the possible cause of the issue. Finding the cause, *e.g.*, weak the signal quality of radio link, the node will decide an action to fix the issue, *e.g.*, enhancing signal quality by applying joint transmission. The EMS with CSON functions enters a SON loop when a regional or global triggering event occurs, so a general policy would be designed to better organize the resource for corresponding slices.

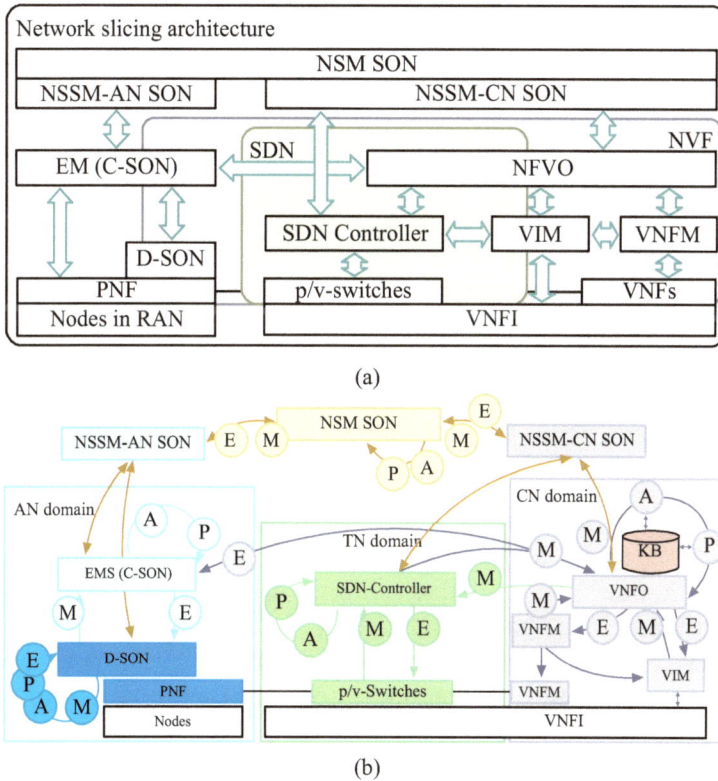

(a)

(b)

Fig. (3). SON-driven network slicing management architecture and automatic management flows. **(a)** The proposed SON-driven NSMA. **(b)** The main automatic management flows with SON control loops identified, where M, A, P, and E represent Monitoring, Analyzing, Planning, and Executing, respectively.

The transport and core network subnets are managed in a centralized way. With the help of the NFV-MANO and SDN Controller, NSSM-CN SON can gather and monitor the global network status information. Assisted by modern deep-learning approaches, NSSM-CN SON can also perform reliable predictions of traffic load and resource utilization of SFCs and generate plans for tuning the resource allocations for SFCs of each slice to achieve dynamic slicing. To address the computational complexity, the analyzing and optimizing functions in NSSM-CN SON can be implemented in the form of a deep reinforcement learning framework. The reinforcement learning provides a general framework to iteratively retrieve (near) optimal solutions to gain new experience, while the accompanied deep neural network keeps the experience for quick reference in the future. Dynamic slicing is very likely to adjust the operating network slicing instances; however, it is not expected the resource allocation changes took place dramatically for slice instances on a large scale. In this article, considering the practical need and constraints, we identify three dynamic resource management approaches for network slicing.

Prediction-based Robust Dynamic Slicing

According to the number of involved entities, three approaches to implementing dynamic slicing in VNF and SFC level, namely scaling, remapping, and appending, are presented. Scaling is a key feature of NFV that permits VNFs to dynamically share pooled resources and usually happens in VNF granularity. Remapping changes a part of the SFC instance to alternative places. Appending attaches a new SFC instance to a target network slice instance. As different situations may occur during the automatic slicing process, a series of algorithms have to be developed to solve VNFs' mapping and resource allocation of NSIs. More explanations are in the sequel.

1) Scaling

When a VNF reaches its capacity limit, there are two ways to extend the capacity. Horizontal scaling refers to adding a new VNF instance to share the workload, and it permits the new VNF instance running on the same or a neighbor node for the same workload. So, it has a lower requirement on resource and capacity of physical nodes. However, as the mapping between VNF and physical node changes, horizontal slicing requires high network topology planning capability. Vertical scaling refers to aggregating more resources for the target VNF, rather than adding a new VNF instance, so the mapping between VNF and the physical node does not change. This means vertical expansion raises higher requirements on resource amount in the physical node and physical resource virtualization technologies. Dynamic scaling is investigated in a few recent papers [22 - 24]. They share a similar algorithm framework which is to iteratively check the NSIs to identify whether the resource demand of an NSI is larger than a predefined threshold. If yes, resources can be allocated to the NSI on demand (vertical scaling); otherwise, new VNFs will be added for the NSI (horizontal scaling), and a new route has to be planned for the newly added VNFs. The differences lie in the prediction approaches they utilized for more precise resource demand perception.

2) Remapping

There are also cases when VNFs in a physical node need to be migrated to another node. A typical case is migrating from a failed physical node to another functioning one. Several VNFs in the same NSI may be simultaneously affected. In this case, re-plan the NFV mapping and resource allocation are needed, so VNFM will inform NFVO to perform the re-plan. During the process, NFVO will consider the attached service level agreement, real-time network condition, and resource occupation and allocation of other NSIs. NFVO may decide to re-generate the slice or only map part of the VNFs according to the actual situation and a predefined preference. If the slice is regenerated, the remapping problem is equivalent to the initial deployment of the slice. If only a part of the VNFs is involved, the remapping problem is equivalent to the problem of slice instance deployment with partially determined variable constraints. Remapping can be found studied in a variety of research related to robustness-aware resource allocation of network services. A common network resource allocation framework for network slice instances is given in [25], where network resources are allocated for each slice instance periodically to gain QoS-awareness for services with different priorities while fully utilize network resources. In [26], a backup-route--based algorithm is proposed to achieve robust network service routing when links or nodes of the network are in failure. It contains one sub-algorithm for finding backup routes for network services in advance and another sub-algorithm for selecting the optimal replacement route according to the practical situations.

3) Appending

When a tenant's traffic increases dramatically, NFVO can establish a new SFC attached to the tenant's NSI and offloading partition of the total traffic to the new SFC. In this way, uninterrupted service for an existing business in the NSI can be guaranteed. Besides, with traffic characterization enabled, NFVO can automatically split a flow with slightly different characteristics into a new SFC to better shape the traffic in SFCs. When it is detected that traffic returns to normal, NFVO can release the newly added SFC and re-farm resources for other network services. This approach gains great flexibility for network slicing, and at the same time provides considerable reliability. According to the depiction, the appending approach mainly contains three components. First, the data flow of network services is classified by a classifier as in [27] to aggregate network data flows with similar properties into the same class. Then, traffic predictors as in [28] can be utilized to give some predictions on the classified data flow, and the services with fluctuating data flow are split into a new slice instance. After that, network resources should be coordinated to serve the newly added network slice instances with the existing ones.

CASE STUDY AND NUMERICAL EVALUATIONS

Case Study

Fig. **4** illustrates a use case reflecting the changes of resource allocation for smart city applications in dynamic network slice instances. In a residential community, a security assurance company (SAC) subscribes a network slice from a network operator for linking its widely deployed security video cameras. The slice has computational resources, so the SAC deploys intelligent video data analysis functions, *e.g.* intrusion detection function (IDF) in the slice to provide automatic security warning service for its customers. In normal cases, IDF will be implemented in an edge server, so video data can be faster processed, and fewer data will go across the network. However, when a sudden cutoff of electricity or gas occurs, wearable augmented reality devices can help repairers from vertical company to perform maintenance. To achieve real-time environment augmented, a temporal network slice will be established for this vertical special task (VST), and computational complex augmented reality computing functions are expected to run as close to the scene as possible for lower service delay. Both slices can share edge node resources. Due to the limited resources in edge node and much lower latency requirement of augmented reality services, when the sign of node resource shortage is detected, *e.g.* more repairers joining the maintenance, VNFs of IDF would be remapped into further distributed or center cloud nodes, so precious edge resources can be saved for the VST. Similar situations can be easily observed in other SC-IoT scenarios, and dynamic network slicing enables network re-farm resources for delay-sensitive slices from non-delay-sensitive slices, to better serve slices with the same network resources. In the sequel, we give a preliminary result to show the effects of our design.

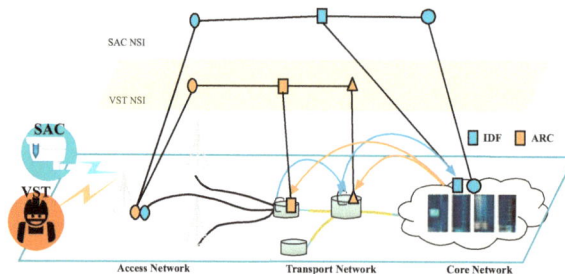

Fig. (4). A illustrative case of automatic network slicing.

Numerical Results

The performance of the scheme is evaluated by simulation in a network scenario, where the numbers of nodes in substrate access, aggregation, and core network are 10, 5, and 3, respectively, while the network capacities are 40, 80, and 80 Gbps. In the network, 100 NSIs are deployed, including 20 for delay-sensitive slice instances and 80 for non-delay-sensitive ones. The simulations are carried out in the MATLAB on the hardware platform, a PC with a 2.1GHz CPU, 16G Memory, 2080Ti GPU. We set the DS traffic increasing from x1 to x7, while non-dela-
-sensitive ones are unchanged.

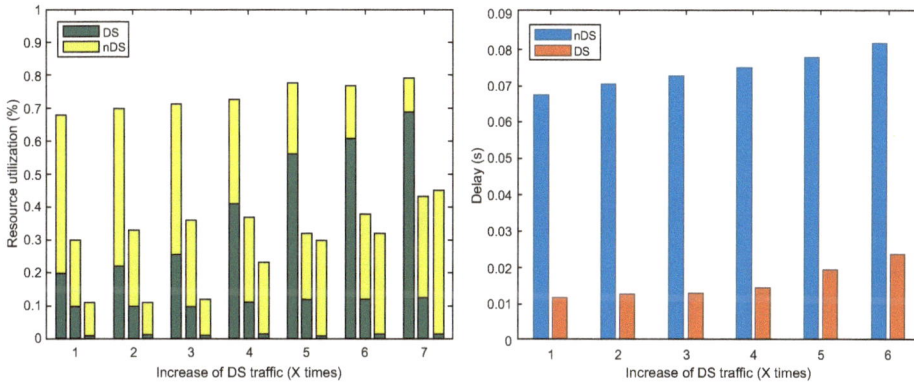

Fig. (5). Averaged network resource utilization rate and latency.

Fig. **5** presents the averaged network resource utilization rate and latency of both delay-sensitive and non-delay sensitive services. In Fig. (**5a**), the bars in each group represent access, transport, and core network domains, respectively. Here we mainly focus on edge node resources in the access network, bandwidth resources in the transport network, and cloud node resources in the core network. We can observe as delay-sensitive traffic increases, more edge resources in the access network domain will be drawn back from non-delay-sensitive network services to serve the delay-sensitive ones, so more and more edge resources are occupied by delay-sensitive services. The total utilization rate of edge resources grows up a little, due to the total traffic increase. In the transport and core networks, the thing is different, the increase of delay-sensitive services, will not make delay-sensitive services occupy more resources in transport and core networks; however, non-delay-sensitive will use more node resources in cloud nodes. Sending more traffic to the core network, non-delay-sensitive will use more transport network bandwidth resources, which is also reflected in Fig. (**5b**). The delay of delay-sensitive services will increase continuously, while that of delay-sensitive services stay almost unchanged. But both of them are within the requirement. With these traffic-aware dynamics, delay-sensitive services will be served with a low-latency guarantee, while non-delay-sensitive services will also get served with satisfying delay performance.

SUMMARY

This chapter investigates the intelligent network slicing management and control technology for future 6G networks. In particular, we investigated some major issues that need to be resolved when slicing is deployed in a network, such as a network service function mapping and multi-tenant slicing provision. Other than existing works in this area, we employ a SON-based framework to handle network and radio resource slicing for IoT applications. Simulation results illustrate that our approach can provide flexible and reliable answers to complicated scenarios.

It should be noted that there is still a long way before a practical automatic network slicing mechanism comes true. For now, this chapter just presents our main idea on automatic network slicing and the provided high-level framework could incorporate other detailed algorithms and mechanisms to release the power of self-organizing. We hope the chapter can inspire more potential work on this promising research direction.

CONSENT FOR PUBLICATION

Not applicable.

CONFLICT OF INTEREST

The author declares no conflict of interest, financial or otherwise.

ACKNOWLEDGEMENTS

Declared none.

REFERENCES

[1]   H. Sun, C. Wang, and B. I. Ahmad, *From Internet of Things to Smart Cities - Enabling Technologies* CRC Press: Taylor & Francis Group, 2017.

[2]   X. An, C. Zhou, R. Trivisonno, R. Guerzoni, A. Kaloxylos, D. Soldani, and A. Hecker, "On end to end network slicing for 6G communication systems", *Trans. Emerg. Telecommun. Technol.,* vol. 28, no. 4, 2017.
[http://dx.doi.org/10.1002/ett.3058]

[3]   M. Richart, J. Baliosian, J. Serrat, and J-L. Gorricho, "Resource slicing in virtual wireless networks: A survey", *IEEE eTrans. Netw. Serv. Manag.,* vol. 13, no. 3, pp. 462-476, 2016.
[http://dx.doi.org/10.1109/TNSM.2016.2597295]

[4]   M.R. Raza, M. Fiorani, A. Rostami, P. Ohlen, L. Wosinska, and P. Monti, "Dynamic Slicing Approach

for Multi-Tenant 6G Transport Networks", *J. Opt. Commun. Netw.,* vol. 10, no. 1, p. A77, 2018.
[http://dx.doi.org/10.1364/JOCN.10.000A77]

[5]   J. Ordonez-Lucena, P. Ameigeiras, D. Lopez, J.J. Ramos-Munoz, J. Lorca, and J. Folgueira, "Network slicing for 6G with sdn/nfv: Concepts, architectures, and challenges", *IEEE Commun. Mag.,* vol. 55, no. 5, pp. 80-87, 2017.
[http://dx.doi.org/10.1109/MCOM.2017.1600935]

[6]   J. Sallent, "Perez-Romero, R. Ferrus, and R. Agusti, "On radio access network slicing from a radio resource management perspective", *IEEE Wirel. Commun.,* vol. 24, no. 5, pp. 166-174, 2017.
[http://dx.doi.org/10.1109/MWC.2017.1600220WC]

[7]   Q. Li, R.Q. Hu, Y. Qian, and G. Wu, "Cooperative communications for wireless networks: techniques and applications in lte-advanced systems", *IEEE Wirel. Commun.,* vol. 19, no. 2, 2012.

[8]   X. Foukas, G. Patounas, A. Elmokashfi, and M.K. Marina, "Network slicing in 6G: Survey and challenges", *IEEE Commun. Mag.,* vol. 55, no. 5, pp. 94-100, 2017.
[http://dx.doi.org/10.1109/MCOM.2017.1600951]

[9]   Y. Jia, H. Tian, S. Fan, P. Zhao, and K. Zhao, "Bankruptcy game based resource allocation algorithm for 6G Cloud-RAN slicing", *IEEE Wireless Communications and Networking Conference (WCNC),* pp. 1-6, 2018.

[10]   X. Li, R. Casellas, G. Landi, A. de la Oliva, and X. Costa-Perez, "A. GarciaSaavedra, T. Deiss, L. Cominardi, and R. Vilalta, "6G-crosshaul network slicing: Enabling multi-tenancy in mobile transport networks", *IEEE Commun. Mag.,* vol. 55, no. 8, pp. 128-137, 2017.
[http://dx.doi.org/10.1109/MCOM.2017.1600921]

[11]   S. Vassilaras, L. Gkatzikis, N. Liakopoulos, I.N. Stiakogiannakis, M. Qi, L. Shi, L. Liu, M. Debbah, and G.S. Paschos, "The algorithmic aspects of network slicing", *IEEE Commun. Mag.,* vol. 55, no. 8, pp. 112-119, 2017.
[http://dx.doi.org/10.1109/MCOM.2017.1600939]

[12]   Q. Jia, R. Xie, T. Huang, J. Liu, and Y. Liu, "Efficient caching resource allocation for network slicing in 6G core network", *IET Commun.,* vol. 11, no. 18, pp. 2792-2799, 2017.
[http://dx.doi.org/10.1049/iet-com.2017.0539]

[13]   L. Gong, H. Jiang, Y. Wang, and Z. Zhu, "Novel location-constrained virtual network embedding lc-vne algorithms towards integrated node and link mapping", *IEEE/ACM Trans. Netw.,* vol. 24, no. 6, pp. 3648-3661, 2016.
[http://dx.doi.org/10.1109/TNET.2016.2533625]

[14]   L.U. Khan, I. Yaqoob, N.H. Tran, Z. Han, and C.S. Hong, "Network Slicing: Recent Advances, Taxonomy, Requirements, and Open Research Challenges", *IEEE Access,* vol. 8, pp. 36009-36028, 2020.
[http://dx.doi.org/10.1109/ACCESS.2020.2975072]

[15]   A. Devlic, "NESMO: Network slicing management and orchestration framework", *IEEE International Conference on Communications Workshops (ICC Workshops), Paris,* 2017.

[16]   W. Saad, M. Bennis, and M. Chen, "A Vision of 6G Wireless Systems: Applications, Trends, Technologies, and Open Research Problems", *IEEE Netw.,* vol. 34, no. 3, pp. 134-142, 2020.
[http://dx.doi.org/10.1109/MNET.001.1900287]

[17]   S.D. Alfoudi, S.H.S. Newaz, A. Otebolaku, G.M. Lee, and R. Pereira, "An Efficient Resource Management Mechanism for Network Slicing in a LTE Network", *IEEE Access,* vol. 7, pp. 89441-89457, 2019.
[http://dx.doi.org/10.1109/ACCESS.2019.2926446]

[18]   T. Afolabi, "Network Slicing and Softwarization: A Survey on Principles, Enabling Technologies, and Solutions", *IEEE Communications Surveys & Tutorials,* vol. 20, no. 3, pp. 2429-2453, 2018.

[19]   X. Shen, "AI-Assisted Network-Slicing Based Next-Generation Wireless Networks", *in IEEE Open*

Journal of Vehicular Technology, vol. 1, pp. 45-66, 2020.
[http://dx.doi.org/10.1109/OJVT.2020.2965100]

[20] S. Redana, A. Kaloxylos, A. Galis, P. Rost, and V. Jungnickel, "View on 5G architecture", *5G-PPP Architecture WG, 5GPPP, Germany, White Paper,* 2016.

[21] J. Zhou, W. Zhao, and S. Chen, "Dynamic Network Slice Scaling Assisted by Prediction in 5G Network", *IEEE Access,* vol. 8, pp. 133700-133712, 2020.
[http://dx.doi.org/10.1109/ACCESS.2020.3010623]

[22] M. Bouzid, "Cooperative AI-based e2e Network Slice Scaling",
[http://dx.doi.org/10.1109/INFCOMW.2019.8845198]

[23] J. Nan, "Slice-Scaling Strategy Based on Representation Learning in Flex-Grid Optical Networks", *2019 Optical Fiber Communications Conference and Exhibition (OFC),* 2019pp. 1-3 San Diego, CA, USA
[http://dx.doi.org/10.1364/OFC.2019.M2A.3]

[24] W. Li, Y. Zi, L. Feng, F. Zhou, P. Yu, and X. Qiu, *Latency-Optimal Virtual Network Functions Resource Allocation for 5G Backhaul Transport Network Slicing,* 2019.
[http://dx.doi.org/10.3390/app9040701]

[25] B. Liu, P. Yu, X. Qiu, and L. Shi, "Risk-Aware Service Routes Planning for System Protection Communication Networks of Software-Defined Networking in Energy Internet", *IEEE Access,* vol. 8, pp. 91005-91019, 2020.
[http://dx.doi.org/10.1109/ACCESS.2020.2993362]

[26] M. Lopez-Martin, B. Carro, A. Sanchez-Esguevillas, and J. Lloret, "Network Traffic Classifier with Convolutional and Recurrent Neural Networks for Internet of Things", *IEEE Access,* vol. 5, pp. 18042-18050, 2017.
[http://dx.doi.org/10.1109/ACCESS.2017.2747560]

[27] R. Vinayakumar, K.P. Soman, and P. Poornachandran, "Applying deep learning approaches for network traffic prediction", *2017 International Conference on Advances in Computing, Communications and Informatics (ICACCI),* 2017, pp. 2353–2358.
[http://dx.doi.org/10.1109/ICACCI.2017.8126198]

[28] V. Sciancalepore, K. Samdanis, X. Costa-Perez, D. Bega, M. Gramaglia, and A. Banchs, *Mobile traffic forecasting for maximizing 5G network slicing resource utilization,* 2017.
[http://dx.doi.org/10.1109/INFOCOM.2017.8057230]

CHAPTER 6

Applications and Implementations of 6G Internet of Things

Tao Hong[1,*] and **Fei Qi**[2]

[1] *School of Electronic and Information Engineering, Beihang University, Beijing, China*

[2] *China Telecom Beijing Research Institute, Beijing, China*

Abstract: The Internet of things (IoT) has been the information infrastructure of a digitalized society and drives the newest wave of industrial development. With the rise of smart vehicular IoT applications, such as intelligent transport, smart navigation, and automatic driving, vehicular IoT is gaining some new features that cannot be fully addressed by current 5G networks. This chapter presents an overview of the vehicular IoT developing trend and discusses its relationship to 5G and the coming generation. It also presents some survey results from recent literature on the challenges and promising technologies for vehicular massive IoT.

Keywords: Automatic driving, Intelligent transport, Smart navigation, Vehicular IoT.

INTRODUCTION

Internet of Things

IoT is the "Internet of Everything Connected". It is the network that extends and expands on the basis of the internet. It is a huge network formed by combining various information sensing devices with the internet. It can connect people to people and people to things. The internet of everything has become a distinctive feature of 5G, which will profoundly change personal life and economic and social development. At the same time, the rapid development of the IoT will surpass the connection between things and enter a new era of cognitive or intelligent IoT. IoT and 6G will be more deeply integrated to generate a digital twin virtual world. Information can be transmitted between people and people, people and things and things and things in the physical world through the digital world. The twin virtual world is the simulation and prediction of the physical

---

* **Corresponding author Tao Hong:** School of Electronic and Information Engineering, Beihang University, Beijing, China; Tel: 0086 10 82317231; Fax: 0086 10 82338348; E-mail: hongtao1974@163.com

Xianzhong Xie, Bo Rong, Michel Kadoch (Eds.)

world, which accurately reflects and predicts the true state of the physical world. It realizes the "digital twin, intelligent endogenous" to help human beings further liberate themselves and improve the quality of life and the efficiency of production and governance of the entire society. Therefore, the vision of "digital creation of a new world, intelligent communication of all things" is realized.

The rapid development of communication and information technologies is making the massive connections between humans and machines has become a novel feature of communication systems. IoT provides ubiquitous connectivity for anyone and anything (including mobile phones, smart homes, industrial equipment and vehicles, *etc*.) at anytime and anywhere, enabling them to communicate, coordinate, and share information with each other so as to efficiently make decisions and perform their respective tasks. All types of communications, such as machine-to-machine (M2M), human-to-machine (H2M), and human-to-human (H2H), can be carried out in IoT. IoT is a vast area of research,having various forms, complex technologies and broad implications.

In light of the information life cycle of collection, transmission, processing and utilization, a five-layer architecture for IoT is widely adopted, namely, the perception and recognition layer, data communication layer, network interconnection layer, management layer and application layer. These layers perform different functions, but are also closely connected. Below the application layer, various technologies on the same layer are complementary each other to apply to different environments and constitute a full set of strategies for the technologies at this layer. And different layers provide configurations and combinations of various technologies to form a complete solution according to application requirements.

THE IMPACT OF IOT APPLICATION REQUIREMENTS ON 6G

As societal needs continue to evolve, there has been a marked rise in a plethora of emerging use cases that cannot be served satisfactorily with 5G. For example, the next generation of VAR, *i.e.*, holographic teleportation, requires Tbps-level data rates and microsecond-level latency, which cannot be achieved with even the millimeter wave (mmWave) frequency bands within 5G. Further, increasing industrial automation and the move from Industry 4.0 to the upcoming Industry X.0 paradigm will push connectivity density well beyondthe106 km2 metric that 5G is designed for, in addition to requiring an overhaul of existing network management practices. Further, an increase in the connection density will also result in demands for improved energy efficiency, which 5G is not designed for. Consequently, the research community has gravitated towards addressing the aforementioned major challenges, and we posit that ongoing research in the domains of terahertz band communications, intelligent surfaces, and environments, and network automation, for example, may very well hold the key to the future of wireless.

The future development of 6G technology may involve but not limited to the following more important technical fields. These technologies include: (i) a network operating at the THz band with abundant spectrum resources, (ii) intelligent communication environments that enable a wireless propagation environment with active signal transmission and reception, (iii) pervasive artificial intelligence, (iv) large-scale network automation, (v) an all-spectrum reconfigurable front-end for dynamic spectrum access, (vi) ambient backscatter communications for energy savings, (vii) the Internet of Space Things enabled by CubeSats and unmanned aerial vehicles (UAVs), and (viii) cell-free massive MIMO communication networks. We also make note of three very promising technologies that are expected to shape the future of communications, yet will not be sufficiently mature for 6G. These include: (i) the Internet of NanoThings, (ii) the Internet of BioNanoThings, and (iii) quantum communications.

NEW FEATURES OF VEHICULAR IOT APPLICATIONS

During the recent years, the intensive deployment of low-cost wireless sensors with computing, and storing resource gives vehicles powerful information processing abilities to optimize driving decisions and enhance traffic safety [1]. With the improvement of the ability of vehicles to perceive their surroundings and communicate with other roadside utilities, the vehicular network is evolving towards the massive machine type of communication (mMTC) vision. Beyond-5G (B5G) is a promising technology to provide ubiquitous connected vehicle-t--everything (V2X) links, including vehicle-to-vehicle (V2V), vehicle-t--infrastructure (V2I), vehicle-to-network (V2N), and vehicle-to-pedestrian (V2P) communications [2]. Table **1** lists V2X services and restrictions. Besides the traditional massive IoT features, the increasing driving automation demand relies heavily on the safety-critical services that are delivered through vehicular networks; thus, strict requirements of latency, and bandwidth, *etc.* are imposed on vehicular IoT. These requirements are even more harsh owning to the highly dynamic and spatiotemporal complexity of network topologies and fast-varying wireless propagation environment caused by the strong mobility of vehicles [1].

VEHICULAR IOT DEMANDS MORE THAN 5G

To build smart transport system, vehicles need to share a large amount of data to support applications, such as updating real-time road traffic situation and emergency information, referring 3D navigation map for path planning, and even automatic driving [3]. Although a majority of information is processed by artificial intelligent (AI) in local vehicles to support automatic driving, the high level autonomous control and scheduling of traffic flows in the smart transportation system rely heavily on the instance information from massive connected vehicles and roadside units. Since vehicles generally move fast, extremely low round-trip time is a prerequisite for effective information delivery in vehicular networks [3]. In addition, the emerging 3D integrated transportation system, featured by the coexistence of space-air-ground shown in Fig. (**1**) and even underwater vehicles, extremely enlarges the scale and complexity of the vehicular IoT [4].

Table 1. V2X services and restrictions.

Service Category	Safety Service	Autonomous Driving Service	Internet and Infotainment Service
Communication type	V2V, V2P, V2I	V2V, V2I, V2N	V2N
Delay	100 ms	10 ms	no strict restrictions
Reliability	About 99%	99.999%	no strict restrictions
Transmission rate	About Mb/s	10 Mb/s	Browsing: 0.5 Mbps HD video: 15 Mbps

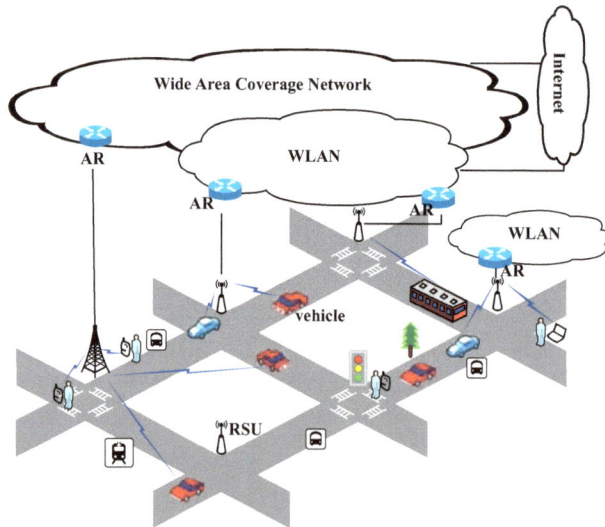

Fig. (1). Architecture of heterogeneous vehicular network in 6G.

In such a context, vehicular IoT should be constructed with better heterogeneous structure to support the data delivery under dynamic networking of the novel network structure. The AI-based self-adaptation enables self-optimization and the fast response to improve quality of service (QoS) and reduce latency, which are fundamental to vehicular networks. Developing an intelligent system is the goal of sixth-generation (6G), enabling the networks to adapt to the dynamic environment to meet different constraints of different applications [4]. Thus, by the end of the 5G developing cycle, we will probably not see the mature of vehicular IoT, as only in B5G/6G can its potential can be fully developed [3].

THE CONCEPT AND VISION OF 6G MASSIVE IOT

The Development Of Massive Iot Concepts

5G Massive IoT

Driven by the increasing informatizing demand and advanced IoT technologies, 10's of IoT billions of devices that will not be intervened by people are envisioned connect to networks in 2020s. To address the emerging market, ITU initialized the solicitation of proposals of 5G standards in 2015 and identified mMTC as one of the three major application scenarios with extreme or even contradictory performance requirements that candidate 5G standards have to address as presented in Fig. (**2**). As a worldwide leading communication standard developing organization, the third generation partnership project (3GPP) launched some standards workgroups to study solutions for mMTC like extended coverage GSM (EC-GSM), LTE for machine-type communication (LTE-M), and narrowband IoT (NB-IoT). 5G mMTC aims to provide massive connection of devices with usually sporadic data transmission [1]. In this context, the latency and bandwidth are not a big concern in 5G massive IoT, while how to optimize power utilization to prolong battery life in these devices for more than 10 years becomes the main concern [2].

The main features of 5G massive IoT, including the massive connectivity are as follows.

- Small data packets, normally 10~20 bytes
- Massive devices (300,000~1 billion devices per cell)
- Usually uplink transmission
- Low data rate, about10 kb/s per user
- Low device complexity and cost
- Sporadic (non-periodic) data
- High energy efficient and long battery life

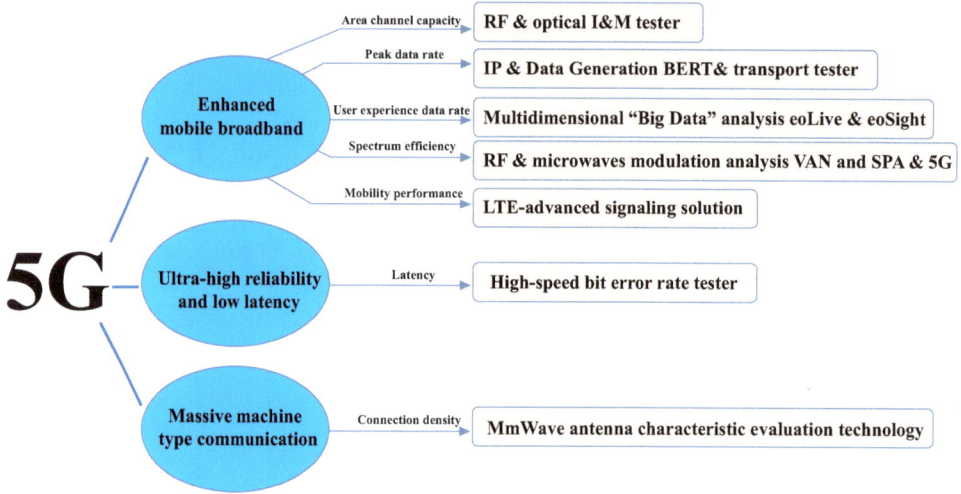

Fig. (2). 5G Performance goals for different application scenarios.

B5G/6G Massive IoT

With the explosive growth of IoT, the number of IoT devices is forecasted to be 41.6 billion and the amount of data generated from them will be 79.4 ZB data by 2025. Thus 6G IoT needs to be designed to satisfy higher requirements in terms of coverage, capacity, and connectivity, *etc*. 6G is expected to have powerful calculation and communication capabilities to support transmitting and processing of mass data generated by IoT devices.

Being the next generation of 5G, 6G is envisioned to support 1000 times more simultaneous connections per unit area than that in 5G. Performance requirements of IoT devices in 6G massive IoT tend to be heterogeneous since they need to deliver increasingly diverse applications. For example, a huge number of digital meters in power, gas, and water supply utilities do not have strict requirements on latency and only need to communicate periodically, while massive actuators and controllers in smart factory for automatic control will have strict requirements on latency and require real-time communication. In addition, 6G massive IoT is expected to serve a variety of industry verticals, including public safety, transportation, environment, healthcare, manufacturing, *etc.*, and thus heterogeneous IoT devices with different attributes and requirements will be utilized in the rich diversity of applications. As all types of communication services are compound together, it will be difficult to identify the distinct characteristics of the typical scenarios as in 5G. Therefore, other than merely enhanced massive connectivity, the key drivers of 6G is widely deemed as the combination of the characteristics in 5G like high data rate, high reliability, low latency, and massive connectivity shown in Fig. (**3**). This is quite different from the case in 5G where those features are use case specified and will not be necessarily supported in the same use case. The trend can be identified from the observations that emerging advanced applications, such as virtual navigation, automatic driving, V2X, bear both the massive connection and URLLC features [5].

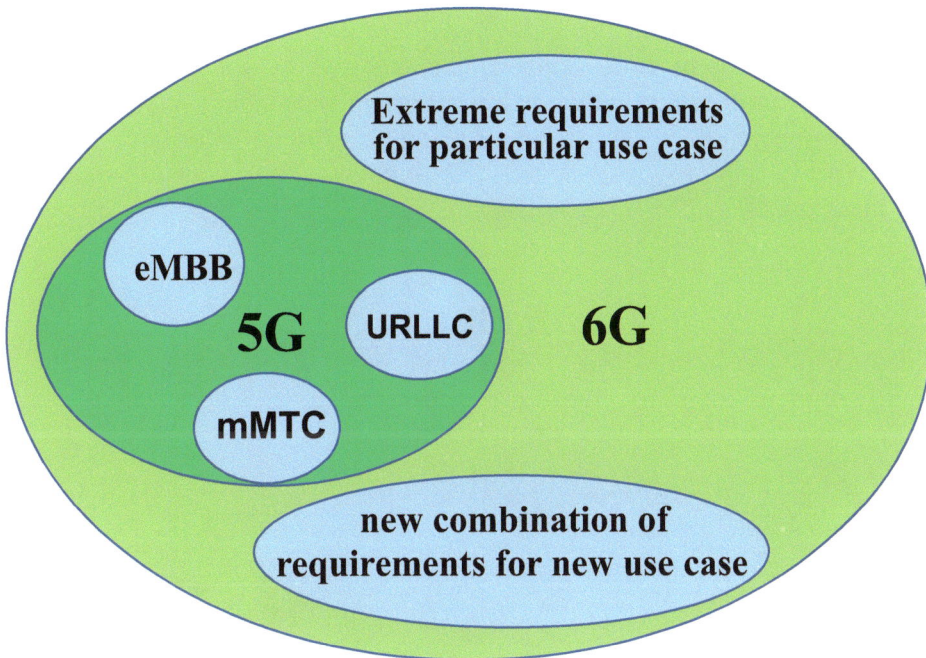

Fig. (3). 6G calls for a combination of all the 5G features.

Main Features of 6G Massive IoT

As novel information and communication services might bring huge profits, operators are racing to build next-generation communication facilities. Among various potential Internet services, many various IoT devices are united to supply seamless transmission service transparently [6]. Therefore, it is generally believed that the IoTs will develop significantly in the next few years. In the future 6G, novel diverse IoT applications with different QoS requirements will appear. For example, V2X communications have high mobility, industrial control requires ultra-high reliability, holographic communication needs ultra-low delay, and health monitoring devices have energy consumption restrictions [7]. However, current network does not have the ability to supply personalized services to different users. In order to support transmission with high data rate, IoT terminals are supposed to access to 6G through the massive heterogeneous base stations (BSs), as Fig. (**4**) shows.

Comprehensive perception, reliable transmission, and intelligent processing are the primary features of IoT. Comprehensive perception means obtaining information of IoT devices by various sensors and other technical methods of perception, capture, and measurement. By the convergence of diverse information networks and the Internet, objects are connected to the information network, and reliable information interaction and sharing are carried out to achieve reliable transmission. Intelligent processing means analyzing and processing data in different region, industries and sectors through intelligent methods such as machine learning and reinforcement learning to provide information about the physical world, economy and society, which carry out decision-making and control intelligently.

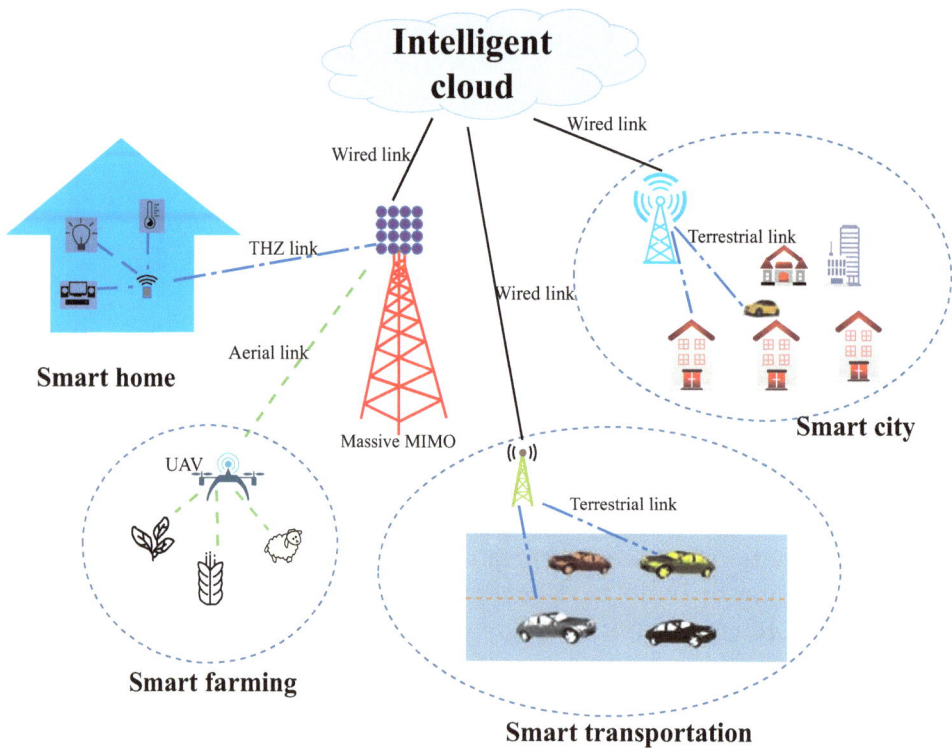

Fig. (4). Intelligent 6G IoT network.

Compared with conventional IoT networks, 6G IoT will have several novel features. Firstly, energy harvesting technology will make up for the disadvantage of short battery life. Different energy sources like solar energy will be used to continuously provide energy for IoT devices. Mobile edge computing (MEC) technology will greatly alleviate the limitations of limited computing capacity. Some computing tasks that require low latency will be offloaded to MEC servers for real-time data processing. Second, short-range bandwidth offensive IoT services will use high-frequency spectrum resources including mmWave and terahertz (THz). Therefore, 6G IoT will have increased peak data rate compared with 5G IoT. For remote IoT services shown in Fig. (4), drones are used to provide instant communication and computing assistance to expand the communication coverage. Finally, AI approaches, such as deep learning and reinforcement learning, are enabling technologies to 6G to promote the intelligentization development.

Internet-connected devices can interact directly to provide an interactive intelligence and environment perception. There are many challenges of IoT, such as random deployment, limited uplink and downlink wireless resources, high mobility of some devices, low battery power and heterogeneity of devices. For numerous IoT devices, centralized resource management is not appropriate because of the high computing and information acquisition cost. The devices connected to IoT are envisaged to be intelligent, which can make decisions and execute actions autonomously. Therefore, distributed resource allocation is necessary to the future IoT. Table **2** lists the main characteristics of 6G networks. 6G and the IoTs bring a great number of end devices, as well as mass data and workload.

Table 2. Main fetures of current and 5G networks and 6G networks.

Current and 5G Networks	6G Networks
AI models are preloaded in edge devices	Edge devices have the ability to execute AI algorithms by themselves
Communication at Gigahertz frequencies	Communication at mmWave and Terahertz frequencies
macrocells	massive base stations and servers in ultra-dense networks
Pre-IoT systems with a small amount of devices	massive connected devices

Typical Scenarios & Applications

With the advent of the IoT era, massive IoT devices need to be connected to the wireless network to provide various applications for industry, agriculture, medicine, transportation and other fields. Although many IoT communications have been deployed so far, large-scale connections and better energy efficiency have not been considered. Large-scale machine communication (MTC) means a huge number of connected devices, such as wearable devices, smart homes, and Internet of vehicles, and low cost need to be ensured. In addition, the IoT covers diverse applications. Real-time monitoring may need wide bandwidth. The data throughput is very small for asset tracking, but inevitably there will be a lot of switching as the object moves. Smart meters only need to transmit small data sporadically. Therefore, no single technology can meet the specific needs of IoT devices.

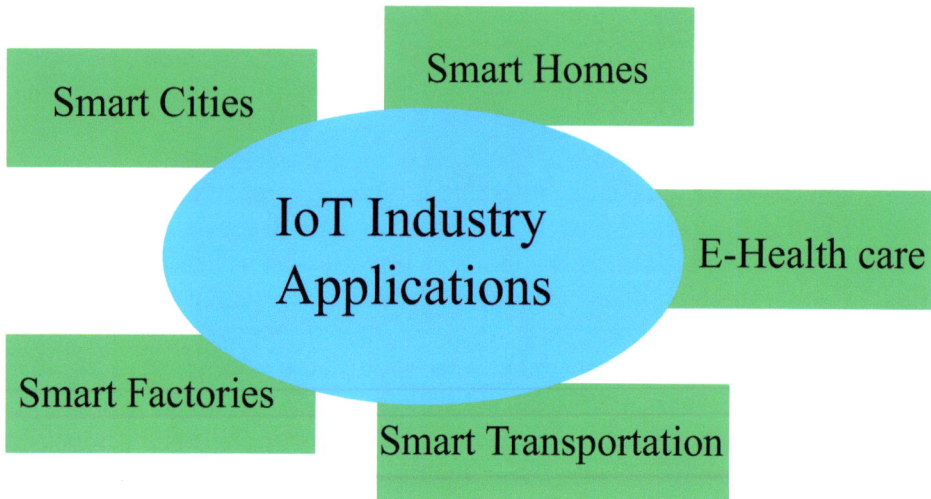

Fig. (5). Applications of 6G IoT.

At present, 5G, IoT, MEC, AI, robotics, blockchain, virtual reality and other technologies are accelerating their integration into the field of industrial IoTs, jointly promoting the arrival of the "fourth industrial revolution". 5G IoT provides rich applications. In the 6G, the communication between machines and devices will not be subject to human interference. Some promising applications of 6G IoT are shown in Fig. (**5**).

1. Smart Cities: The smart cities have a lot of intelligent devices with the ability of information perception and processing. By the intelligent computing and communication, the devices and infrastructure can be monitored or managed. For example, bus arrival information and parking information can be obtained conveniently and timely.
2. Smart Homes: From smart pet feeders to smart thermostats that support network connections, modern homes are equipped with many interconnected devices. In recent years, the market for these products has grown dramatically. According to the digital market outlook by Statista in 2020, it is estimated that

there will be 481 million smart home devices worldwide by 2025. These smart devices enable homeowners to increase efficiency and minimize their carbon footprint.

3. E-Healthcare: Healthcare is an application scenario that needs to be prioritized. Wearable devices can be used for health monitoring and telemedicine can make patients get health service in time. IoT technology and mobile computing technology are utilized in e-healthcare to connect patients, doctors and medical equipment, enhancing the quality of medical services. In the future, more AI technologies will be integrated into the medical industry to make e-healthcare more intelligent.

4. Smart Transportation: Smart transportation is also called Intelligent transportation system (ITS). In the future, ITS will be a combination of monitoring, communication, and control functions. Vehicles will have sensors, intelligent information processing and control modules, and they will be connected to ITS, so that vehicles can be monitored, managed, and informed with traffic conditions. ITS can help vehicles to plan their paths reasonably and avoid traffic accidents as far as possible for the sake of traffic efficiency and safety. Hence, it has received increasing attention from various countries.

5. Smart Factories: Smart factories refers to the factories with intelligence. 6G can provide ultra-high reliability and ultra-low delay communication, providing support for smart factories. A smart factory is a typical scenario for the 6G IoT. Smart factories make use of the IoT, 6G, AI and other advanced technologies of perception, communication, decision making and control to enhance intelligence of factories. Smart factories are able to timely and accurately collect the production line data, monitor and control the production process, which can reduce labor cost and improve productivity.

In the next section, we will focus on vehicular massive IoT and potential technologies.

THE CHALLENGE AND POTENTIAL TECHNOLOGIES OF 6G VEHICULAR IOT

Challenges for Vehicular IoT to be Addressed in 6G

Coexistence and Cooperation of Diverse RATs For Optimized CV2X

To support future vehicular IoT with compound demand on low delay, high transfer rate, and massive connectivity, sufficient wireless is a prerequisite shown in Fig. (**6**). New spectrum resources for vehicular IoT is able to be introduced by utilizing technologies, such as massive multiple input multiple output (MIMO), mmWave bands, VLC, and THz communication. However, how to decide a reasonable methodology and assumptions for optimize the performance of vehicular IoT need further work [8].

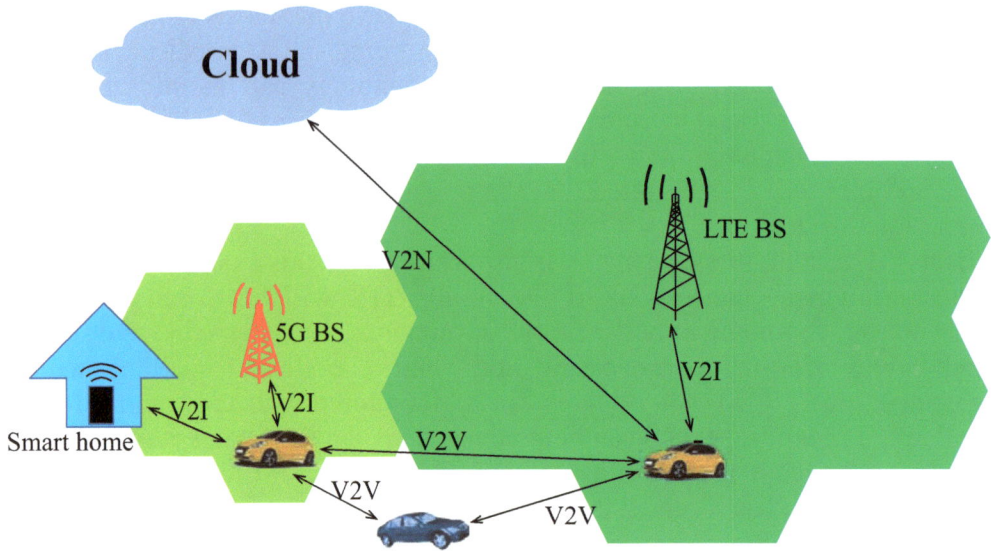

Fig. (6). Multi-RAT for cellular V2X communications.

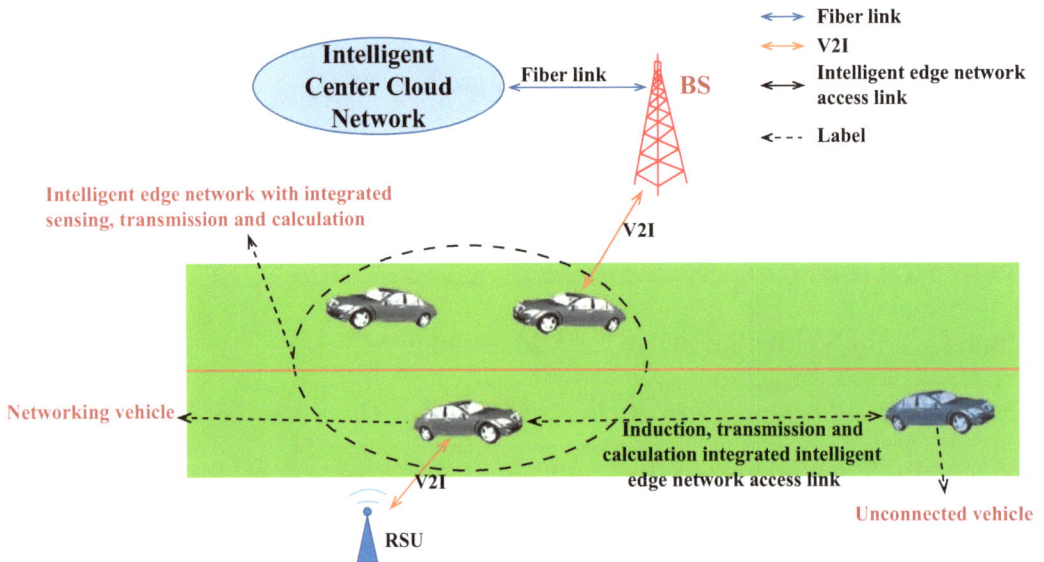

Fig. (7). Illustration of integrated sensing, computing, and communication network.

Convergence of Sensing, Computing, Communication, Caching, and Control

In the advanced applications in vehicular networks, such as high-level automatic driving, the decision making of automatic control relies on sufficiently gathered environment information, while the efficient delivery of the information is on account of the optimization of communication resource allocation. It is prominent that the information infrastructure and facilities, including more than the communication but also computing, caching, and storage, should be fully utilized to fulfill the extreme QoE requirement for the services. Broadly speaking, the convergence of intelligent wireless sensing, communication, computing, caching, and control is needed for 6G vehicular IoT shown in Fig. (**7**), and deserves further investigation [9].

Vehicular IoT Technical Verifications and Testing

A lot of effort has been executed to push forward the research of vehicular IoT all over the world, but it may be difficult to have the C-V2X functionalities and performance evaluated in some use cases in the field testing scenarios, such as emergency cases for automatic driving, rescuing of dropping aerial vehicles. Digital twin provides virtual representations of systems and can serve as a good method for vehicular IoT based autonomous driving test [8].

Promising Technologies of 6G IoT

IoT needs a solution that can provide low power consumption and wide coverage. Generally, it needs to meet the following four points:

1. Low technical cost: support wider deployment and improve the return on investment of applications.
2. Low power consumption: The battery life needs to be very long.
3. Wide coverage: it can be connected to equipment in underground, building and rural environment.
4. High connection capacity: There will be numerous connected IoT devices, which will undoubtedly be a huge load.

In terms of computation, the traditional method of sending first and calculating later does not apply to the mass data in 6G cell IoT, because it has high delay and low spectrum efficiency. In order to solve this problem [10], proposed a very promising solution, namely air computing (AirComp), which can directly calculate the objective function including the sum structure on the wireless multiple access channels (MACs). AirComp combined with MIMO technology can perform spatial multi-channel multi-function calculations and use spatial beamforming to reduce calculation errors in 6G cellular IoT.

As for transmission, in the conventional orthogonal multiple-access (OMA) scheme, orthogonal radio resources in the time, frequency or code domain are allocated to different users, so there is no interference from other users. It is hard to reach the total capacity of a multi-user system by OMA and the amount of orthogonal resources and allocation granularity restrict the number of users. The problems exiting in OMA could be handled by non-orthogonal multiple-access (NOMA), where controllable interference and receiver's tolerance for complexity to increase are allowed. Different users are allowed to use time and frequency resources simultaneously in the same space layer through power domain or code domain multiplexing. Power domain multiplexing is to assign power levels to different users to optimize the system performance. Code domain multiplexing could reach certain expansion gains and shaping gains at the expense of increased channel bandwidth.

So as to achieve accurate calculations and efficient communication at the same time, IoT devices require adequate energy supply. Nevertheless, supplying energy to large-scale IoT is not easy. The large-scale IoTs with frequent battery replacement is not realistic for the consideration of saving labor cost and protecting the environment. Therefore, it is very attractive to use wireless power transmission (WPT) technology to realize one-to-many charging using the openness of wireless broadcast channels. In order to enhance the efficiency of WPT in fading channels, energy beamforming technology is promising. In fact, supplying energy, aggregating data and transmitting information are competing for the same radio resources to improve performance.

Although 6G has just started and is still in the stage of conception, the layout began when 5G was in the ascendant, indicating that everyone has realized the shortcomings of 5G. Therefore, more perfect technologies are needed to promote the development of industries such as AI, IoT, smart driving, and smart manufacturing. And 6G will be more than just a communication technology. It will integrate satellite communications, blockchain, AI, IoT and other high-tech requirements into one to form a huge network of communication without blind spots in the future. At that time, IoT will truly have the foundation of networking and will usher in a real explosive period.

SUMMARY

In this chapter, we investigate the developing trend of vehicular IoT and identify its relationship to 6G communication system. The smart vehicular IoT applications play more and more significant role in modern society. However, most of the features of the applications put forward challenging novel requirements on performance and capability of the information infrastructure. Due to the large number of vehicles, especially when considering the air-groun--underwater compounded future transport system with unmanned control, vehicular IoT bears the massive IoT feature and, at the same time, it demands low latency, and possibly large bandwidth. These are out of the envisioned scope of 5G and will possibly only be implemented in 6G. The chapter also presents some survey results from recent literature of the challenges and promising technologies for vehicular massive IoT.

CONSENT FOR PUBLICATION

Not applicable.

CONFLICT OF INTEREST

The author declares no conflict of interest, financial or otherwise.

ACKNOWLEDGEMENTS

Declared none.

REFERENCES

[1]   C. Kalalas, and J. Alonso-Zarate, *Massive Connectivity in 5G and Beyond: Technical Enablers for the Energy and Automotive Verticals,* 2020.
      [http://dx.doi.org/10.1109/6GSUMMIT49458.2020.9083809]

[2]   M. Shehab, A.K. Hagelskjær, A.E. Kalør, P. Popovski, and H. Alves, "Traffic Prediction Based Fast Uplink Grant for Massive IoT",
      [http://dx.doi.org/10.1109/PIMRC48278.2020.9217258]

[3]   F. Tariq, M.R.A. Khandaker, K. Wong, M.A. Imran, M. Bennis, and M. Debbah, "A Speculative Study on 6G", *IEEE Wirel. Commun.,* vol. 27, no. 4, pp. 118-125, 2020.
      [http://dx.doi.org/10.1109/MWC.001.1900488]

[4]   F. Tang, Y. Kawamoto, N. Kato, and J. Liu, "Future Intelligent and Secure Vehicular Network Toward 6G: Machine-Learning Approaches", *Proc. IEEE,* vol. 108, no. 2, pp. 292-307, 2020.
      [http://dx.doi.org/10.1109/JPROC.2019.2954595]

[5]   "Priority-based initial access for URLLC traffic in massive IoT networks: Schemes and performance analysis",

[6]   B. Mao, Y. Kawamoto, and N. Kato, *AI-Based Joint Optimization of QoS and Security for 6G Energy Harvesting Internet of Things,* 2020.
      [http://dx.doi.org/10.1109/JIOT.2020.2982417]

[7]   J. Li, N. Zhang, Q. Ye, W. Shi, W. Zhuang, and X. Shen, "Joint Resource Allocation and Online Virtual Network Embedding for 5G Networks",
      [http://dx.doi.org/10.1109/GLOCOM.2017.8254072]

[8]   X. You, ">Towards 6G wireless communication networks: Vision, enabling technologies, and new paradigm shifts", *SCIENCE CHINA Information Sciences.*

[9]   Z. Zhang, "6G Wireless Networks: Vision, Requirements, Architecture, and Key Technologies", *IEEE Veh. Technol. Mag.,* vol. 14, no. 3, pp. 28-41, 2019.

[http://dx.doi.org/10.1109/MVT.2019.2921208]

[10]   L. Chen, N. Zhao, Y. Chen, F.R. Yu, and G. Wei, Over-the-Air Computation for IoT Networks: Computing Multiple Functions with Antenna Arrays [http://dx.doi.org/10.1109/JIOT.2018.2843321]

CHAPTER 7

Cloud/edge Computing and Big Data System with 6G

Peng Yu[1,*] and **Lei Shi**[2]

[1] State Key Laboratory of Networking and Switching Technology, Beijing University of Posts and Telecommunications, Beijing, P. R. China

[2] Carlow Institute of Technology, Carlow, Ireland

Abstract: This chapter gives a systematic introduction to cloud/edge computing and big data system. Cloud computing provides the ability to use flexible and telescopic services for cloud users and could implement through various hosted services provided by the Internet. Edge computing refers to the open platform which uses the network, computing, storage, and application core capabilities on the side of the object or data source and provides the nearest end service to avoid the relatively long delay to reach the data cloud center. Big data technology refers to the analysis of potentially useful information from a large number of data. Analysis of big data can get hidden patterns, unknown relevance, customer trends, and other messages to ensure comprehensive data management. In addition, the speed of 6G is faster, and its service field becomes more extensive compared with the previous generation communication technologies, which makes 6G play a more important or extensive role in the future of the technology field and society. In this regard, the authors also analyze the effect of 6G on cloud/edge computing and big data system. According to the future users' demand for 6G and the characteristics of 6G itself, cloud/edge computing and big data system will play an irreplaceable role in achieving high efficiency and benefits.

Keywords: Big data, Cloud computing, Edge computing.

CLOUD COMPUTING WITH 6G

The Development and Characteristics of Cloud Computing

With the number of mobile smartphone users increasing rapidly, more and more users are accessing the Internet *via* mobile phones. Meanwhile, cloud computing affects mobile services by changing the structure of Internet services. Cloud computing is developing every day, which provides a dynamic circumstance of a

---

* **Corresponding author Peng Yu:** State Key Laboratory of Networking and Switching Technology, Beijing University of Posts and Telecommunications, Beijing, P. R. China; Tel: 0086 10 62283225; Fax: 0086 10 62283145; E-mail: yupeng@bupt.edu.cn

technical nature [1] [2]. In this environment, Cloud computing creates innovative solutions and services. In the past three years, enterprises wanted to explore more efficient and valid paths for using their IT investment so that they can adopt cloud computing rapidly. Cloud computing provides the ability to use flexible and telescopic services for cloud users. In consequence of that, users do not have to install the computing resources on their systems. Cloud computing promises that it could provide cheap and flexible services for users. Meanwhile, Cloud computing allows small-scale organizations and individuals to manage services around the world. Nevertheless, although people have been researching a lot in this field, some open challenges still exist. To start with and to run the applications of cloud computing smoothly, the Internet connection must be robust, steady, and rapid. Cloud computing needs a high-speed network and big data handling capacity. However, network resources are not limitless, so running cloud computing normally is inseparable from planning network resources reasonably. In addition, cloud computing is unsafe in applications. The TCP/IP system of the Internet is not safe at present. The process of network applications has some disadvantages, such as spreading viruses or eavesdropping on data. Therefore, it is not completely safe to put all the individual or enterprise data into cloud storage.

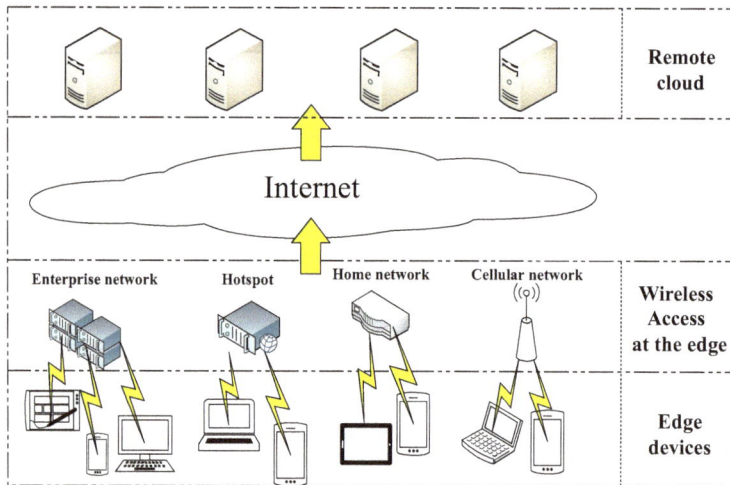

Fig. (1). Concept of Internet edge.

Cloud computing uses the concept of utility computing (Fig. **1**). It could implement through various hosted services provided by the Internet. Over the past decade, it has been developed with extremely fast speed. Its business model is the pay-as-you-go model of metered services as people are familiar with [3]. In this model, users only pay charges for what they use instead of paying for all the things. Meanwhile, this model could meet the extra requirements of services in real-time. Inspired by general low-cost, high-speed Internet, the capability of virtual processing, and the technology of parallel and distributed computing, the idea of cloud computing was proposed.

BASIC CONCEPTS OF CLOUD COMPUTING

Cloud computing is characterized by manageability, scalability, and availability. Besides, cloud computing has a variety of advantages such as convenience, service on demand, versatility, flexibility, stability, *etc*. Three service delivery models are primarily provided by cloud computing: infrastructure as a service (IaaS), platform as a service (PaaS), and software as a service (SaaS). Also, Cloud computing provides four development patterns: public cloud, private cloud, hybrid cloud, community cloud, and virtual private cloud, which is presented in Fig. (**2**).

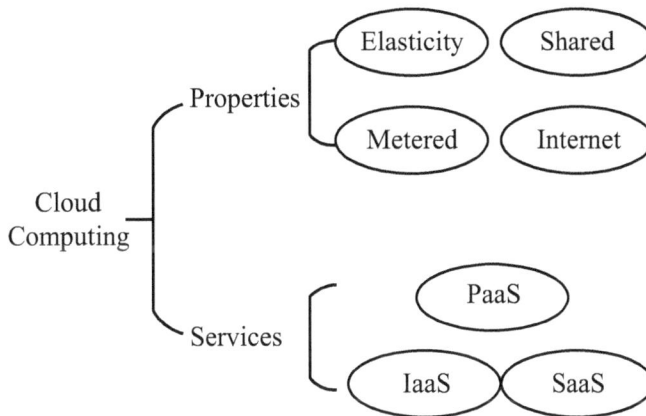

Fig. (2). Cloud computing infrastructure.

Private Cloud

In this kind of development pattern, cloud is owned by a private organization, due to cloud applications aim to serve its own business, information in cloud is only shared within its organization. Cloud applications may be internal or external and could be supervised by a third party or the organization [4]. A private cloud guarantees the performance, reliability and security at the highest level.

Community Cloud

Community cloud means basic facility of cloud is used by several organizations at the same time, and it supports specific community which has common concerns such as security requirements, assignments and so on. Community cloud resembles a private cloud. It has some additional functions and could provide services to the organizations that have similar demand type.

Public Cloud

Public cloud is opposite to private cloud. It is used by the public and owned by a service provider. Service Level Agreement (SLA) is a contract which service between a provider and a consumer. It could specify the quality of public cloud's services, the demands of consumers, and the commitment of providers. Since users could use EC2 (Elastic Cloud Compute) service, which could support virtual machines as computing resources, it has become very popular to provide public cloud service provisioned by VM.

Hybrid Cloud

The cloud infrastructure consists of two or more clouds which could be private, community or public clouds shown in Fig. (**3**). These clouds are tied together by standardized or proprietary technology which could realize the portability of data and applications. Hybrid cloud is the extension of cloud computing with private, public, and community cloud computing techniques. Above-mentioned clouds are integrated together to perform multiple tasks which could cope with demands of private, public and community organizations. Hybrid clouds are more flexible than private and public clouds. Compared with public cloud, they control application data to a higher degree. In addition, the best split between public and private cloud components should be confirmed carefully in the process of designing a hybrid cloud.

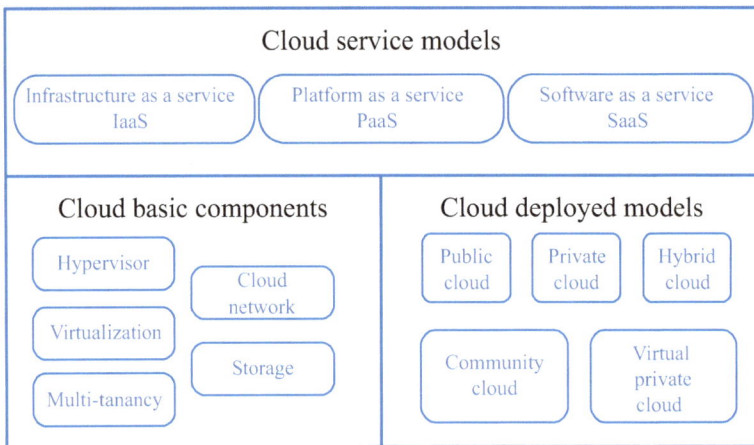

Fig. (3). Cloud computing framework.

NEW APPLICATION OF CLOUD COMPUTING IN 6G ERA

6G has played a key role in mobile communication with the integration of new technologies. The speed of 6G is faster, and the service field of it becomes more extensive compared with the previous generation communication technologies, which makes 6G play a more important or extensive role in the future of technology field and society [5]. In the future 6G era, cloud computing will continue to expand its influence, and the privacy protection will become a important part of applications. The first 6G wireless summit was held in March 2019. Which started the 6Genesis program to execute multiple research directions of 6G. The directions are about wireless connection, the technology of equipment and circuit and distributed intelligent computing. 6G will combine with cloud computing, big data and artificial intelligence (AI) to form the integration of more advanced and ubiquitous information and intelligence.

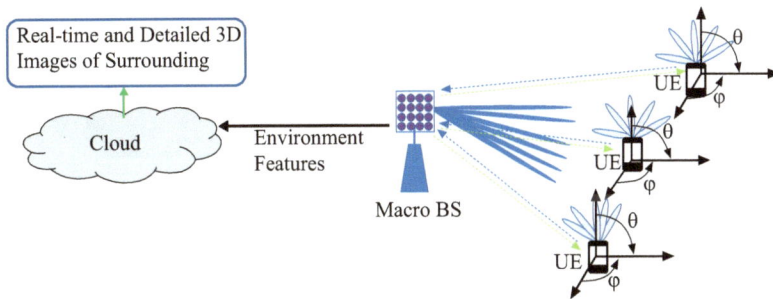

Fig. (4). Illustration of sensing and positioning with mmWave communication.

SENSING AND POSITIONING IN 6G ERA WITH CLOUD

The technology of sensing and positioning using mmWave and THz frequencies can measure the environment surround and acquire the features about the environment based on the observation results of the signal sent to the sensor shown in Fig. (4), which take advantage of the wavelength and frequency selectivity. The environment features are transmitted to the could platform, which could be processed in real time. It is apparent that the sensing and positioning technology is suitable for the indoor scene with the help of mmWave and THz communication systems. The volume of the environment data produced by the equipment is huge. This large amount of data generated by Sensing application needs to be processed and stored by cloud server. Sensing applications can generate maps or scene graphs of any location. This technology can generate real-time and detailed 3D images of the surrounding, or the world, which will be uploaded/shared to the cloud by future devices. In addition, certain materials possess specific frequency vibration absorption throughout the THz band, which makes the detection of the presence of certain substances based on Frequency scanning spectroscopy become possible.

In the technology of sensing and positioning with THz frequencies in the 6G scenario, positioning is to find the position of a fixed or mobile user based on the known position of a base station or other kind of reference. For the purpose of accurate positioning, cloud computing can perform complex reflection calculations on positioning technology. In a non-line-of-sight (NLOS) environment, light cannot reach the base stations (BSs) directly due to specular reflections from walls and metal surfaces, resulting in a decrease in accuracy. Therefore, the positioning applications are likely to be able to scanning the surrounding and generate detailed three-dimensional maps in real time, or acquire them from the cloud. So that Non-direct targets can be "observed" by scanning the environment quickly to identify all reflective obstacles in the current environment.

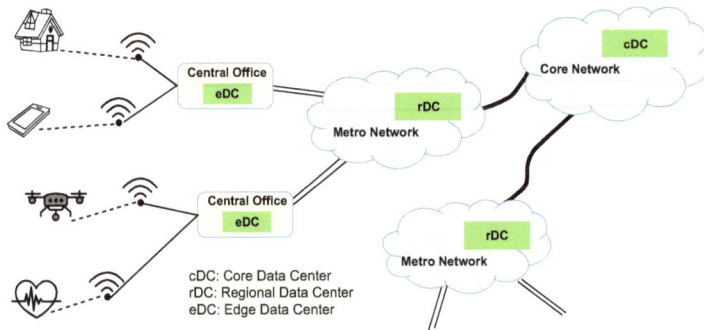

Fig. (5). Slicing with slice-specific software optimization.

THE PLACEMENT OF SLICE-SPECIFIC FUNCTION IN CLOUD

In addition to extending the traditional connectivity architecture to various subnets and multi-connection scenarios, it can also expect to make further progress in network slicing and virtualization [6]. Network slicing can produce separate software stack by highly specialized slicing, as shown in Fig. (5). Virous applications, such as mart home devices, UE, unmanned aerial vehicle (UAV) and intelligent medical devices, are widely connected to the core network with the central offices and metro networks, which could promote the integration of various wireless communication modes. With the traditional access methods to realize slices, we can also allow connectionless access based on the low-throughput IoT slices. In gateway devices, relay devices, edge and regional clouds and many other devices can be utilized to accomplish the needs of specific slices, by which a flexibly slice-specific function placement can be formed.

THE CONTRIBUTION OF CLOUD COMPUTING FOR MOBILE COMMUNICATIONS IN 6G

For the coming of 6G, we must pay more efforts in developing the foundations of artificial intelligence (AI) algorithms, big data capabilities and new computing architectures shown in Fig. (6), in which cloud computing will make a great contribution. Among these three important fields, AI provides the intelligence to all devices and applications with the information provided by big data and the computational power of cloud computing. Big data is a key role in the combination of the technologies [7]. Without it, there is no resource for AI to perform its remarkable learning algorithms and cloud computing will have no computing objects. The Integration of AI, big data and cloud computing provides 6G strong data understanding ability.

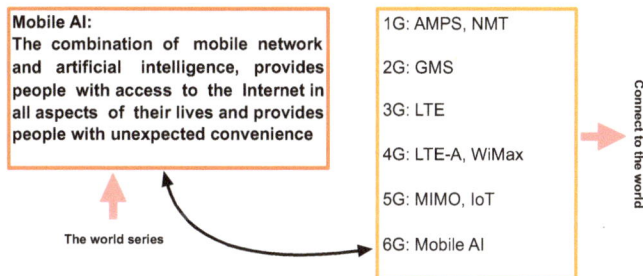

Mobile AI:
The combination of mobile network and artificial intelligence, provides people with access to the Internet in all aspects of their lives and provides people with unexpected convenience

The world series

1G: AMPS, NMT

2G: GMS

3G: LTE

4G: LTE-A, WiMax

5G: MIMO, IoT

6G: Mobile AI

Connect to the world

Fig. (6). The major goals of wireless mobile systems from 1G to 6G.

APPLICATION OF EDGE COMPUTING

The Emergence of MEC

In order to adapt to the explosive growth of large-scale mobile data in the future and accelerate the development of new business applications, mobile communication system has started to change. Mobile communication system will no longer be a pure communication framework shown in Fig. (7), but developed into a service provider based on the close combination of information transmission, processing and control. In the future, the network will not only provide services for mobile terminal users, but also face many different vertical industries, such as industrial Internet, Internet of things, automatic driving, remote surgery, *etc*. These new services have different characteristics, and the requirements of communication system are also different. But overall: they usually require very low end-to-end latency to process large amounts of data. However, the existing infrastructure and communication network cannot meet the requirements of new services. At present, due to the limited computing power of end users, a large number of resource demand tasks that cannot be performed locally will be transmitted to the data cloud center through the communication network for centralized processing. However, a large amount of data in the transmission process is bound to generate additional delay, which will significantly reduce the performance of real-time interactive applications [8 - 10]. The emergence of the first mock exam has led to the emergence of the edge computing model. The edge computing originated in the media field. It refers to the open platform which uses the network, computing, storage and application core capabilities on the side of the object or data source, and provides the nearest end service to avoid the relatively long delay to reach the data cloud center. This model has been accepted by academia and industry.

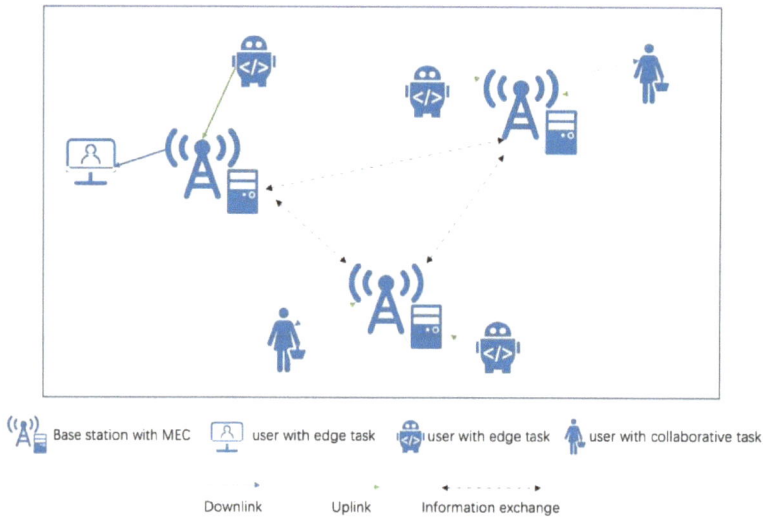

Fig. (7). The architecture of an MEC based network.

In the deployment of MEC technologies, the major challenge is the security issue. Enterprise security requirements for network security: business application security, user data security, network security, stability and reliability. The security risks and challenges MEC faces come from the increased risks of infrastructure security and the difficulty of data security protection. MEC hardware infrastructure deployed on the edge of the network is more vulnerable to external attackers. Attacked edge computing devices extend the risk to the network infrastructure. The distance between the attacker and the edge computing physical device is shortened, which brings security risks such as physical device destruction, service interruption, user privacy and data leakage. It is easier for malicious users to illegally listen or tamper with traffic. At present, the industry is also considering some corresponding solutions improving the capability of MEC services through distributed intrusion detection technology, virtual machine isolation technology, *etc*. In addition, the application of MEC can save energy for users, where more tasks can be completed by MEC server with the grid powering.

The Advantages and Disadvantages of MEC

The Advantages of MEC

With the rapid development of wireless communication systems and resource allocation, especially when the fifth generation (5G) network is developed, the development of the Internet of Things (IoT) has made great progress, which at the same time provides strong support for the development of smart cities. Two key criteria for evaluating IoT performance are latency and cost. Moreover, mobile edge computing (MEC) deploys computing resources to edge servers close to end users, achieving a good balance between cost efficiency and low-latency access, making it an attractive new choice. In the IoT system, nodes can not only get in touch with each other but also have the ability to store data and compute. In the Internet of Things system, wireless caching technology is very important, because it can greatly help improve the quality of user experience. Fortunately, due to the development of storage technology, storage costs have dropped rapidly. As everyone knows, with the exception of wireless caching technology, mobile edge computing (MEC) technology also plays a very important role in IoT systems.

Nowadays, the demand for delay-sensitive services is rapidly increasing to facilitate the support of key tasks such as industrial IoT, autonomous vehicles, e-health care, *etc*. To solve this problem, many feasible solutions can meet the stringent requirements of service delay including MEC. Specifically, MEC places the server near its users, which effectively reduces the propagation delay. The services of telecom operators are naturally covered in 5G cellular networks across the country, and these nodes can complete computing tasks with the assistance of nearby nodes instead of remote clouds. However, it is very complicated to determine the location of the MEC server and associate users with the server. In this way, waiting time and energy consumption can be significantly reduced. This model reduces the physical distance between computing and storage resources and wireless terminal equipment. MECs provide cloud computing services at the edge of the radio access network (RAN), allowing access to computing and storage resources close to end user equipment (UE). Customer requests are processed locally at the edge of the network instead of forwarding traffic to the mobile backhaul, thereby reducing the load on the mobile backhaul and avoiding network bottlenecks. In addition, being close to end users helps support the strict latency requirements of 5G services and use cases. At the same time, the ever-increasing computing power and storage capacity of mobile terminal devices are valuable resources that can be used to enhance MEC.

The Disadvantages of MEC

Today, cloud computing still has extraordinary computing power and storage advantages, and most data still needs to be transmitted to the cloud for calculation. However, mobile edge computing also has some difficult problems to be broken through. With the development of science and technology and the increase in the number of devices, the amount of data that needs to be transmitted is also increasing. The application of deep models exceeds the computing power of MEC devices. At the same time, data exchange is also required between MEC devices. There is a complicated division of labor and cooperation mode between MEC devices, and fast message transmission between devices is required to ensure complete the task smoothly.

Edge Computing in 6G Era

Concept and Vision of 6G

The 6G network is a key force leading the development of information and intelligence in 2030, and it is expected to provide better performance than 5G in emerging services and other applications. The high-speed transmission rate of 6G meets the conditions and requirements of most intelligent networks, so it is widely used. 6G proposes a large autonomous network architecture that integrates space, air, ground, and underwater networks, and provides wireless connectivity without any limitations. In addition, we also discussed artificial intelligence (AI) and machine learning, and related design and innovation of autonomous network air interfaces. The key factor for 6G is the convergence of all past features such as network density, high throughput, high reliability and large-scale connectivity. Therefore, the 6G system will continue the trend of previous generations on the premise of new technology and new services. These new services include artificial intelligence, smart wearable devices, implants, self-driving cars, computational reality devices, sensors and three-dimensional maps. The processing of large amounts of data and the high-speed connection of devices are the most important requirements of 6G wireless network. Finally, we identified a number of promising 6G ecosystems shown in Fig. (**8**), including Terahertz (THz) communications, very large antenna arrays, large Intelligent Surfaces (LISS) and holographic beamforming (HBF), nano-IoT and MEC.

Effect of 6G on MEC

With a faster transmission speed of 6G, it can solve the above problem better. 5G is the highest technical level in the field of communication at present, its network can reach a speed of 1G/ min, while 6G is predicted to break through 10G/s, and its theoretical speed can even reach 1t/s. In the 6G era, the peak rate index of single terminal of 6G is predicted to reach 100Gbps, and the total throughput of single cell is predicted to reach 1Tbps. Meanwhile, low delay communication is expected to be mainly concentrated between machines to replace the traditional wired transmission, such as the industrial Internet scene, *etc*. The delay demand should be at the sub-ms level, and the 6G delay index can be predicted to be 0.1ms. Hence, the function of MEC in the 6G era should be further enhanced, and it may even merge the base station with MEC to become a super base station. At the same time, artificial intelligence (AI) will be introduced and some AI-based applications will be introduced in the MEC server side, making the MEC computing power become extremely powerful. According to the analysis and calculation of some data, the MEC intelligent computing capacity in the 6G era is predicted to be 1,000 Tops.

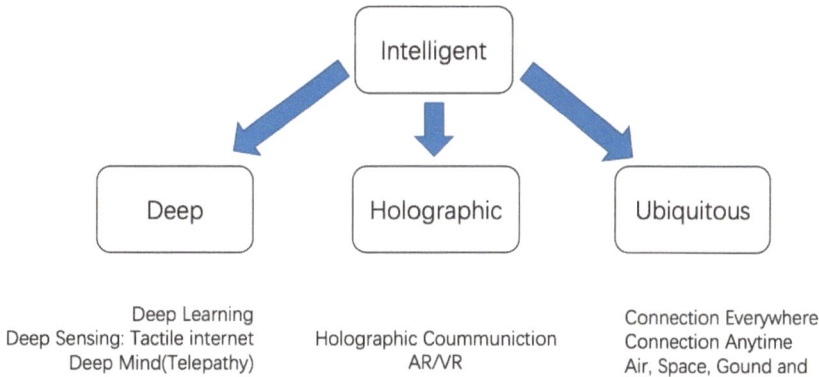

Fig. (8). 6G Vision.

In order to overcome the limitations of 5G development, a unique sixth generation (6G) wireless system was developed. In the field of radio, a concept 6G(sub-) Terahertz radio is expected to be available around 2028, the air interface technology and RF/spectrum problem will be discussed in more detail, the network core technology in the 2023-2025 to zero-contact SDNFV core development, all 2030 target achieve local and even the user-centric micro network support. Network technology is further discussed in [11]. In IoT devices, energy-harvesting technology is expected to stop replacing batteries, and even biodegradable devices could be available within a decade as printing technology matures. We believe that edge computing will play an even bigger role in the 6G era.

From the perspective of the continued development of eMBB, VR business should have the greatest demand for bandwidth in the 6G era. It is expected that after 10 years, VR business will be mature and VR will become a typical user experience business. According to the preset extreme VR experience demand in VR360, bandwidth demand = video resolution × color depth × frame number/compression rate = $(23040 \times 11520) \times 12 \times 120/350 = 1$Gbps, and the typical bandwidth demand for obtaining extreme VR is 1Gbps. Therefore, the recommended user experience rate for 6G is predicted to be at least 1Gbps. Therefore, the major research issue is how to improve the QoE of VR user with the help of MEC cooperation. Mobility is the most basic performance index of mobile communication system. In the 6G era, considering that the high-speed mobility scenario is expected to have a speed of 1200 kilometers per hour, the major research issue is the MEC service mechanism under the high-speed mobility scenario.

MEC with Machine Learning

Overview of ML Model

The advent of artificial intelligence (AI) has revolutionized every industry and has the potential to be the cornerstone of the fourth industrial revolution [12]. The ML model is a computing system used to learn that the characteristics of a system cannot be represented by an explicit mathematical model, and it is often used to implement tasks such as classification, regression, environment interaction, and so on. At present, artificial intelligence has made great progress in computer vision, games, language and other fields. Once the system has completed model training and learned the characteristics of the system, it can use computation to effectively complete the task. Significant advances in artificial intelligence have led to its ever-expanding range of applications and ubiquity. But although ai has a promising future in the cellular field, its development is still in its infancy.

The Value of MEC for AI Model

The rapid development of society has resulted in the frequent use of mobile applications and Internet of Things applications, thus generating a large amount of data. However, it is unrealistic to send all the generated data to the cloud. The recent research results of fog computing and multi-access edge computing (MEC) provide cloud computing storage for data and realize the expansion of network resources shown in Fig. (**9**). These resources are geographically distributed at the edge of the network, closer to mobile and Internet of Things devices, allowing low latency, high bandwidth, and location-based awareness [13]. The future of cellular networks is to leverage the capabilities of air interfaces and cross-layer information to operate artificial intelligence models that enable newer end-user application scenarios. The disadvantage is that multimedia real-time applications require strong computing power, which makes it impossible for some mobile devices to execute these applications. To solve this problem, mobile edge computing (MEC) servers need to be deployed on the wireless base station (BSs) to complete the computing tasks. However, even when MEC servers are deployed through wireless networks, MEC server deployment optimization, task allocation, energy efficiency optimization and other challenges will be faced.

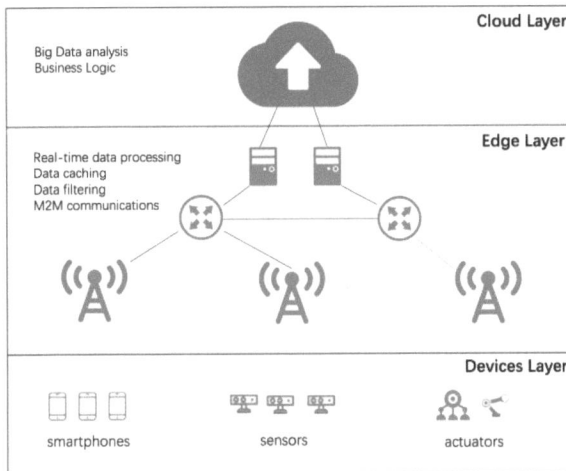

Fig. (9). Edge Computing Architecture.

BIG DATA IN 6G

Overview of Big Data

At present, the importance of big data is self-evident. Big data is not only widely used in the wired field shown in Fig. (**10**), but also appears in people's vision. With the continuous development of technology, big data is growing exponentially, and the authorities have changed their views on big data because of this situation. Big data has become an indispensable part of improving social operation efficiency.

Big data technology refers to the analysis of potentially useful information from a large amount of data. Analysis of big data can get hidden patterns, unknown relevance and customer trends and other messages to ensure comprehensive data management. Big data can be collected from wireless communication and wireless network, including not only historical data and current data, but also data of mobile users, equipment manufacturers and operators, indoor and outdoor experiment data. There are many sources of big data collection. According to different business, the data extracted from video, image and audio can be used as the source of data.

Big data has three characteristics: capacity, speed and diversity. First, the amount of data storage increases exponentially with the diversity of data formats. Nowadays, with the rapid development of social networks, we can find various formats of data on social network service channels, including video, music and image files. The level of data volume is usually TB or Pb. Secondly, big data has strong timeliness. Due to the high speed and change rate of real-time analysis, users always pay attention to the latest updates, which makes the old messages quickly replaced by new ones. Finally, big data can be stored in a variety of formats, such as text, video, SMS, PDF, *etc.*, and it also supports the storage in user-defined format.

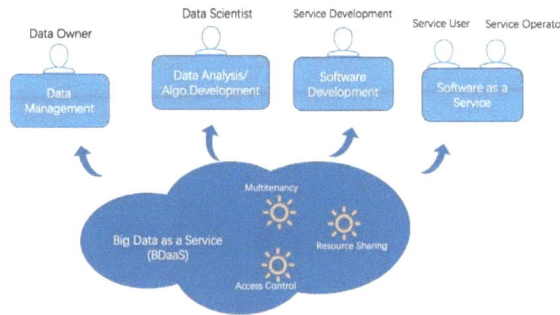

Fig. (10). The application of big data.

Recent Technologies Related to Big Data

Internet of Things

With the emerging of IOT, more and more popular smart devices are used in our daily life, machines embedded with various sensors are connected to each other *via* the Internet. Due to the realization of the Internet of Things (IoT) and the Industrial Internet of Things, drones, sensors and other devices will generate large amounts of data. In the real world, sensors can be used in traffic monitoring, geographic data monitoring, agricultural data collection, weather forecast monitoring, public facilities and household appliances. In most IoT applications, the security of data and equipment becomes the most critical issue. In order to store, access and use sensor data better, the demand of privacy protection and security technologies is becoming stronger. The Internet of Things (IoT) and 6G mobile communication technologies have paved the way for the smart industry and completed industrial optimization and automation well.

Artificial Intelligence Powered by Big Data and 6G

6G and AI are two important partners for our future life, it's believed that AI will make 6G even more powerful. Artificial intelligence technology can provide intelligence for wireless networks through learning and training from big data; therefore, artificial intelligence will be the most innovative technology for designing 6G autonomous networks [14, 15]. Artificial intelligence can also make complex 6G networks more intelligent. The most important effect is that it can achieve high-level efficient network management and optimization, such as network environment awareness, state prediction, active configuration, and dynamic AI will gradually migrate from the cloud to the network edge and adopt multiple layers of AI.

AI plays an important role in 6G networks. Recently, people are increasingly interested in AI. This method is a reasonable choice for detecting malicious behavior of remote entities. There are other requirements that require "intelligence" in the network, such as self-configuration or management complexity. Artificial intelligence depends on abundant computing power. 6G will use ever-increasing computing power to cope with higher bit rates, while gaining more flexibility. In order to meet the needs of various emerging applications, the future sixth-generation (6G) mobile network is expected to become a naturally intelligent, highly dynamic, ultra-dense heterogeneous network, which combines everything with ultra-low latency and high-speed data transmission interconnection. It is believed that artificial intelligence (AI) will be the most innovative technology which can realize intelligent automated network operation, management and maintenance in the future more complex than 6G network.

Role and Applications of Big Data in 6G

The Role of Big Data in Promoting The Development of 6G

According to the future users' demand for 6G and the characteristics of 6G itself, big data will play an irreplaceable role in achieving high efficiency and benefits. In the field of 6G, big data is widely used to process massive data to ensure QoS of 6G architecture, and it can also be used to analyze and determine the best information transmission mode between users, so as to achieve the effect of significantly reducing delay [16]. In the era of explosive growth of data, big data will promote 6G network through automation and automatic optimization. At present, big data has been applied to various fields, such as agriculture, medical treatment, public administration, scientific research, artificial intelligence and so on.

Big Data Applications in 6G Era

Tensor-Computing Based on Big Data

Big data is also applicable to tensor calculation. In the future 6G system architecture, the key to tensor calculation lies in fundamentally improving the computing efficiency of the system, that is, processing the spectrum tensor and the system tensor into a single unit instead of multiple distribution items [17]. In order to realize this core idea, it is necessary to construct spectrum tensor and system tensor from two aspects: spectrum big data and multidimensional system big data. In this process, the operation will be completed by detecting and estimating the filling spectrum big data and system big data. Even after the data is complete, the spectrum and system tensors should be sparse.

Context-Aware Communication

At present, many types of database (*e.g.*, relative database, graphic database) are widely applied to many situations, they can be used to store user specific data registers (such as NR system structure of the unified data management entity), can also be used for the process and calculation on the edge, fog, clouds of advanced technology [18]. As shown in Fig. (**11**), the UE is the processing center of environment awareness for 6G, which has the main sensing functions, such as on-device learning, environment sensing, distributed machine learning and full-frequency radio access. The huge amount of information poses a high challenge to the UE processing capacity and transmission rate [19]. In order to populate and update the database in real time, a crowdsourcing approach must be used to obtain information shown in Fig. (**11**). Therefore, in order to meet this requirement, a standardized interface must be defined. The amount and variety of information expected requires typical custom processing for big data solutions.

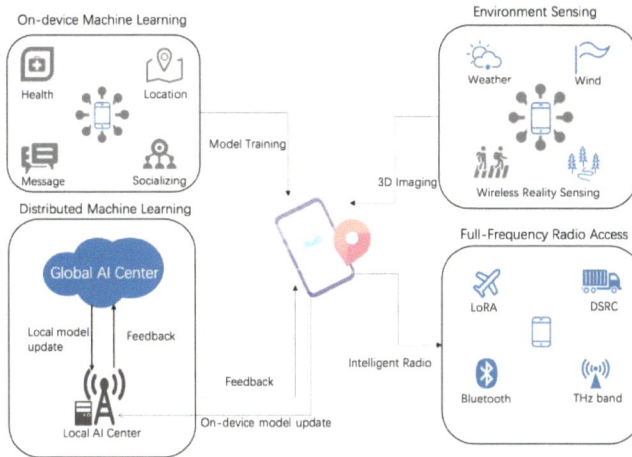

Fig. (11). Environment awareness through 6G and big data.

SUMMARY

This chapter mainly investigates the applications of cloud/edge computing and big data in 6G mobile networks. Firstly, we introduce the basics of cloud computing, including the concepts of private cloud, community cloud, public cloud and hybrid cloud. Secondly, we study the applications of mobile edge computing based on the objective evaluation of advantages and disadvantages of the MEC. It is worth noting that the AI technology, such as deep machine learning, is helpful to improve the performance of 6G MEC system. Thirdly, we give a discussion on 6G big data, which is powered by IoT and AI considerably. The role of big data in 6G is predicted based on the tensor-computing and context-aware communication.

CONSENT FOR PUBLICATION

Not applicable.

CONFLICT OF INTEREST

The author declares no conflict of interest, financial or otherwise.

ACKNOWLEDGEMENTS

Declared none.

REFERENCES

[1]   P. Zhang, and Z. Yan, "A QoS-aware system for mobile cloud computing", *2011 IEEE International Conference on Cloud Computing and Intelligence Systems,* Beijing Beijing, pp. 518-522, 2011. [http://dx.doi.org/10.1109/CCIS.2011.6045122]

[2]   Y. Amanatullah, C. Lim, H.P. Ipung, and A. Juliandri, "Toward cloud computing reference

architecture: Cloud service management perspective", *International Conference on ICT for Smart Society,* Jakarta, pp. 1-4, 2013.
[http://dx.doi.org/10.1109/ICTSS.2013.6588059]

[3]   D. Ardagna, "Cloud and Multi-cloud Computing: Current Challenges and Future Applications", *2015 IEEE/ACM 7th International Workshop on Principles of Engineering Service-Oriented and Cloud Systems,* Florence, pp. 1-2, 2015.
[http://dx.doi.org/10.1109/PESOS.2015.8]

[4]   S. Olariu, "A Survey of Vehicular Cloud Research: Trends, Applications and Challenges", *IEEE Trans. Intell. Transp. Syst.,* vol. 21, no. 6, pp. 2648-2663, 2020.
[http://dx.doi.org/10.1109/TITS.2019.2959743]

[5]   P.J. Sun, "Security and privacy protection in cloud computing: Discussions and challenges", *J. Netw. Comput. Appl.,* vol. 160, p. 102642, 2020.
[http://dx.doi.org/10.1016/j.jnca.2020.102642]

[6]   S. Bera, S. Misra, and J.J.P.C. Rodrigues, "Cloud Computing Applications for Smart Grid: A Survey", *IEEE Trans. Parallel Distrib. Syst.,* vol. 26, no. 5, pp. 1477-1494, 2015.
[http://dx.doi.org/10.1109/TPDS.2014.2321378]

[7]   S. Sakr, A. Liu, D.M. Batista, and M. Alomari, "A Survey of Large Scale Data Management Approaches in Cloud Environments", *IEEE Comm. Surv. and Tutor.,* vol. 13, no. 3, pp. 311-336, 2011.
[http://dx.doi.org/10.1109/SURV.2011.032211.00087]

[8]   J. Wu, Z. Cao, Y. Zhang, and X. Zhang, "Edge-Cloud Collaborative Computation Offloading Model Based on Improved Partical Swarm Optimization in MEC",
[http://dx.doi.org/10.1109/ICPADS47876.2019.00144]

[9]   M. Mehrabi, D. You, V. Latzko, H. Salah, M. Reisslein, and F.H.P. Fitzek, "Device-Enhanced MEC: Multi-Access Edge Computing (MEC) Aided by End Device Computation and Caching: A Survey", *IEEE Access,* vol. 7, pp. 166079-166108, 2019.
[http://dx.doi.org/10.1109/ACCESS.2019.2953172]

[10]   S. Lee, "Low Cost MEC Server Placement and Association in 5G Networks", *2019 International Conference on Information and Communication Technology Convergence (ICTC), Korea (South),* 2019pp. 879-882

[11]   C. Min, "Park, J. Lee, J. Cho and H. Kim, "Issues on supporting public cloud virtual machine provisioning and orchestration", *13th International Conference on Advanced Communication Technology (ICACT2011),* 2011pp. 270-273 Seoul

[12]   Md. Mostafa Zaman Chowdhury, "Shahjalal, Shakil Ahmed and Yeong Min Jang, "6G Wireless Communication Systems: Applications, Requirements, Technologies, Challenges, and Research Directions", *IEEE Open Journal of the Communications Society,* vol. 1, pp. 957-975, 2020.
[http://dx.doi.org/10.1109/OJCOMS.2020.3010270]

[13]   S. Wang, T. Sun, H. Yang, X. Duan, and L. Lu, *6G Network: Towards a Distributed and Autonomous System*, 2020.

[14]   R. Zhao, X. Wang, J. Xia, and L. Fan, "Deep Reinforcement Learning Based Mobile Edge Computing for Intelligent Internet of Things", *Phys. Commun.,* vol. 43, p. 101184, 2020.
[http://dx.doi.org/10.1016/j.phycom.2020.101184]

[15]   R. Shafin, L. Liu, V. Chandrasekhar, H. Chen, J. Reed, and J.C. Zhang, "Artificial Intelligence-Enabled Cellular Networks: A Critical Path to Beyond-5G and 6G", *IEEE Wirel. Commun.,* vol. 27, no. 2, pp. 212-217, 2020.
[http://dx.doi.org/10.1109/MWC.001.1900323]

[16]   Kliks, "Beyond 5G: Big Data Processing for Better Spectrum Utilization", *IEEE Veh. Technol. Mag.,* vol. 15, no. 3, pp. 40-50, 2020.
[http://dx.doi.org/10.1109/MVT.2020.2988415]

[17]   W. Zhang, J. Wu, and C. Wang, "Tensor-computing-based Spectrum Usage Framework for 6G", *ICC 2020-2020 IEEE International Conference on Communications (ICC),,* 2020 Dublin, Ireland, pp. 1-6, 2020.

[18]   G. Liu, Y. Huang, N. Li, J. Dong, J. Jin, Q. Wang, and N. Li, "Vision, requirements and network architecture of 6G mobile network beyond 2030", *China Commun.,* vol. 17, no. 9, pp. 92-104, 2020.
[http://dx.doi.org/10.23919/JCC.2020.09.008]

[19]   S. Wang, M. Chen, X. Liu, C. Yin, S. Cui, and H.V. Poor, "A Machine Learning Approach for Task and Resource Allocation in Mobile Edge Computing Based Networks", *IEEE Internet of Things Journal,* vol. 8, no. 3, pp. 1358-1372, 2021.
[http://dx.doi.org/10.1109/JIOT.2020.3011286]

SUBJECT INDEX